Published by the British Museum (Natural
History), London and the Press Syndicate of the
University of Cambridge
The Pitt Building, Trumpington Street,
Cambridge CB2 1RP
32 East 57th Street, New York, NY 10022, USA
296 Beaconsfield Parade, Middle Park, Melbourne
3206, Australia

First published 1981

**British Library Cataloguing
in Publication Data**

Origin of Species
  1. Evolution
  I. British Museum (Natural History)
  575   QH366.2   80-42170

ISBN 0-521 23878 1 hard covers
ISBN 0-521 28276 4 paperback
Printed in Great Britain by Balding & Mansell, Wisbech, Cambs.

ORIGIN
OF SPECIES

British Museum (Natural History)
Cambridge University Press

Contents

Preface

Every living thing belongs to a species – every plant, every animal, every fungus, every microorganism. But what exactly is a species? And how are new species formed? There are no simple answers to either of these questions, and the origin of species is as hotly debated today as it was when Charles Darwin first published his theory of evolution by natural selection more than 120 years ago.

What is natural selection? How does it work? As readers of this book will discover, natural selection is essentially a very simple idea, and evidence for natural selection can be found amongst living species today. Darwin based his theory of natural selection on four important observations about species – reproductive potential, the effects of the environment, variation and inheritance. **Origin of species** introduces these observations one by one, using a range of examples from the natural world. It then shows how they can be linked to produce the theory of natural

selection, and goes on to examine the role that natural selection might play in the formation of new species.

Throughout the book, the emphasis is on living species, and no attempt is made to reconstruct a history of life on Earth. Moreover, the treatment of genetics is confined to what is needed for a basic understanding of variation and inheritance.

This book is a companion to the major new exhibition **Origin of species**, which will open at the Natural History Museum as part of our centenary celebrations in 1981. Like the exhibition, the book has been planned with the help of experts from the Museum's Scientific Departments, and I should like to take this opportunity of thanking all the people, both within the Museum and outside, who have been involved in its preparation.

R H HEDLEY   Director
British Museum (Natural History)
November 1980

eeping willowHowler monkey Paradise flycatcher Earwig Komodo dragon Fir Toad
l Siamang Secretary bird Stag beetle Crocodile Oak Death cap Spectacled bear Skua A
eba Boomslang Potato Cycad Giant panda Robin Dragonfly Dab Snake's head fritillary I
t Ocelot Shoebill Praying mantis Giant tortoise Broad bean Horsetail Lion Eagle Ladyb
erhead shark Tiger Swan Wasp Midwife toad Cannonball tree Hairy-nosed wombat Ib
d-eating spider Mudskipper Gamboge tree Yapok Emu Blue swimming crab Electric e
lodendron Kangaroo Oven bird Ant Piranha Giraffe Sea squirt Mistletoe Hippopotam
-Dik Bird of paradise Slug Flying fish Rotifer Strangler fig Puffball Killer whale Cucko
llouse Stickleback Breadfruit Orang utan Blue-footed booby Octopus Rattlesnake Ver
uck-billed platypus Penguin Red-backed funnelweb spider Cod Frankincense Ginkg
de-striped jackal Flatworm Seahorse Baobab Zebra Humming bird Bath sponge Salm
odder Stinkhorn Yak Common loon Stingray Flax Armadillo Kiwi Sea anemone Pawpa
nhorn Pelican Jellyfish Axolotl Rosebay willowherb Stag's horn fern Pig Eider duck Be
pin Holly Giant redwood Slow loris Kob Stump-tailed skink Char Lotus Reindeer moss
y Marsupial mole Jay Velvet worm Chameleon Larkspur Monkey-puzzle tree Spiny an
ater Chiffchaff Oyster Fire salamander Tutsan Tenrec Mosquito Coelacanth Snapdrag
amel Sparrow Gooseberry sawfly Frog Spring beauty Horse Wren Desert locust Caim
ose Bladderwrack Badger Osprey Moon moth Rainbow boa Creeping bent Albatross L
et Gecko Asarabacca Stonewort Jaguar Puffin Organ-pipe coral Guppy Liverwort Sund
w Moth Adder Sheep Orache Floating crystalwort Shield bug Carp Olm Otter Thrush L
ewing Venus's flytrap Newt Comb jelly Dolphin Water avens Fanworm Hare Pitcher pl
t Pronghorn Slow-worm Mouse Limpet Hammerhead shark Elephant hawk moth Cha
erhead shark Tiger Swan Wasp Midwife toad Cannonball tree Hairy-nosed wombat Ib
d-eating spider Mudskipper Gamboge tree Yapok Emu Blue swimming crab Electric e
lodendron Kangaroo Oven bird Ant Piranha Giraffe Sea squirt Mistletoe Hippopotam
-Dik Bird of paradise Slug Flying fish Rotifer Strangler fig Puffball Killer whale Cucko
llouse Stickleback Breadfruit Orang utan Blue-footed booby Octopus Rattlesnake Pel
uck-billed platypus Penguin Red-backed funnelweb spider Cod Frankincense Ginkg
de-striped jackal Flatworm Seahorse Baobab Zebra Humming bird Bath sponge Salm
odder Stinkhorn Yak Common loon Stingray Flax Armadillo Kiwi Sea anemone Pawpa
nhorn Pelican Jellyfish Axolotl Rosebay willowherb Stag's horn fern Pig Eider duck Be
apin Holly Giant redwood Slow loris Kob Stump-tailed skink Char Lotus Reindeer mo
y Marsupial mole Jay Velvet worm Chameleon Larkspur Monkey-puzzle tree Spiny an
ater Chiffchaff Oyster Fire salamander Tutsan Tenrec Mosquito Coelacanth Snapdrag
w Moth Adder Sheep Orache Floating crystalwort Shield bug Carp Olm Otter Thrush L

Introduction

Why are there so many different kinds of living things?

One view is that all living things were created just as we see them today, and have never changed. Another view is that the living things we see today have all **evolved** from some distant ancestor by a process of gradual change. But how can evolution have occurred? How could one species have changed into another? Just over 120 years ago, Charles Darwin thought of a way, and called it **natural selection**.

Ever since it was first published, Darwin's theory of natural selection has been one of the most discussed, disputed and misunderstood theories of science. Yet natural selection is a very simple idea, and we can find evidence of natural selection in action in the natural world today.

A corner of the study at Down House, where Darwin worked on his theory of natural selection

Chapter 1

The problem Darwin solved

Even in the middle of the 19th century, evolution was not a new idea. It had been discussed for centuries. But no one could explain how evolution occurred, so the idea was rejected.

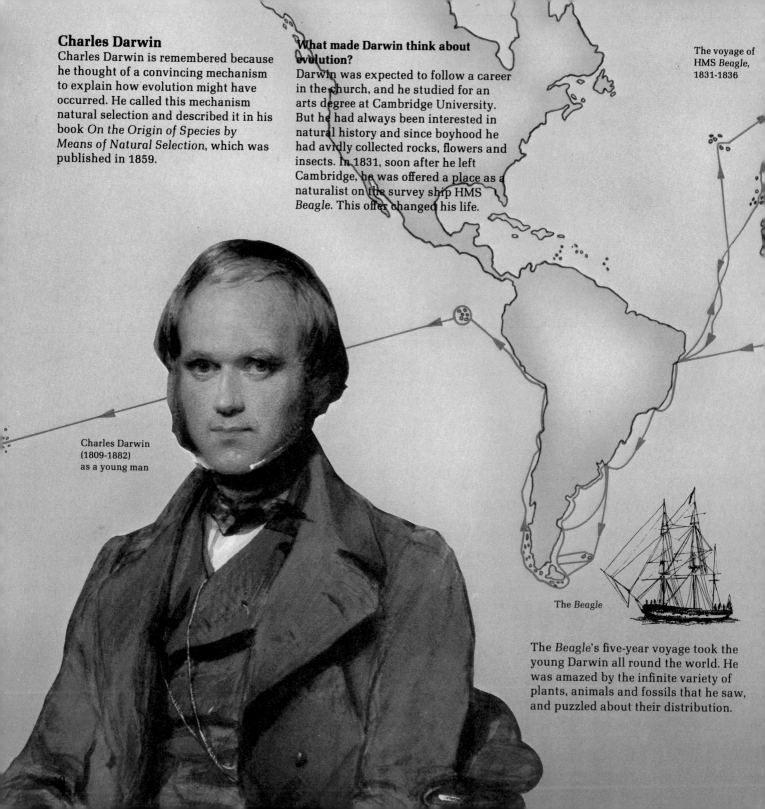

Charles Darwin

Charles Darwin is remembered because he thought of a convincing mechanism to explain how evolution might have occurred. He called this mechanism natural selection and described it in his book *On the Origin of Species by Means of Natural Selection*, which was published in 1859.

What made Darwin think about evolution?

Darwin was expected to follow a career in the church, and he studied for an arts degree at Cambridge University. But he had always been interested in natural history and since boyhood he had avidly collected rocks, flowers and insects. In 1831, soon after he left Cambridge, he was offered a place as a naturalist on the survey ship HMS *Beagle*. This offer changed his life.

The voyage of HMS *Beagle*, 1831-1836

Charles Darwin (1809-1882) as a young man

The *Beagle*

The *Beagle*'s five-year voyage took the young Darwin all round the world. He was amazed by the infinite variety of plants, animals and fossils that he saw, and puzzled about their distribution.

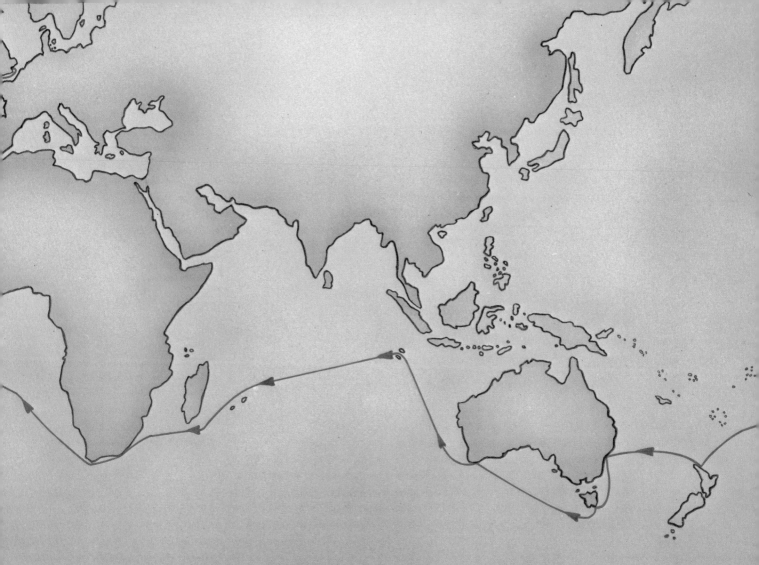

When Darwin returned to England, he began to study the enormous collections of plants and animals he had made during the voyage. He soon began to develop the idea of natural selection, but he was reluctant to publish the theory until he had perfected it. He spent the next twenty years working on his theory and gathering as much evidence as he could.

In 1858, Darwin's studies were interrupted by the arrival of a manuscript from Alfred Wallace, a naturalist working in Indonesia. This manuscript showed that Wallace had independently arrived at the same theory. Darwin and Wallace presented the theory jointly to the scientific world in London, and the following year Darwin published *On the Origin of Species*. The book sold out the day it was published, and caused a storm of controversy because it challenged the accepted views of the time.

Down House in Kent, Darwin's home from 1842–1882

A clue to Darwin's theory

Domesticated plants and animals provided Darwin with an important clue to the way evolution might have occurred.

By breeding together selected individuals, breeders are able to change the characteristics of domesticated plants and animals.

Dachshunds, for example, have changed a lot over the last hundred years or so, as breeders have preferred sleeker, more lightly-built animals with shorter legs and more elegant heads.

A completely new breed of dog, the Staffordshire bull terrier, was produced by breeding together bulldogs and terriers. From each litter of puppies, breeders selected the ones that had the characteristics they wanted. These animals were then bred together and the process was repeated, generation after generation, until the breeders had produced the kind of dog they wanted. You can see that the Staffordshire bull terrier combines some of the qualities of both bulldogs and terriers.

1875

1925

1975

bulldog

terrier

Staffordshire
bull terrier

12

Darwin knew that **domestic selection** over many generations had produced great variety in certain kinds of plants and animals. He thought this might provide a clue to a much more complex process of selection that had been taking place in nature for many thousands of years. He called this process **natural selection**, and thought that it could account for the great diversity of plants and animals he had seen on his travels.

In the following chapters you can find out more about Darwin's theory of evolution by natural selection, and how it fits in with what we observe in nature.

Chapter 2

How we recognize species

You don't have to travel as far as Darwin did to be amazed at the variety of living things.
Just look around you ...

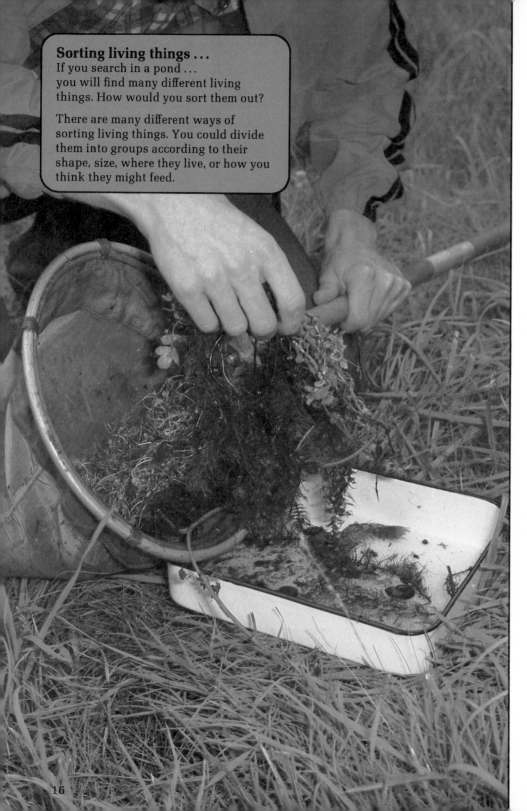

Recognizing species

If you want to recognize different
sorts of living things, you need to put
together the ones that have particular
features in common.

You could start by dividing your
collection into two groups – those that
can move, and those that can't.

The ones that can move are **animals**,
and the ones that cannot move are
probably **plants**.

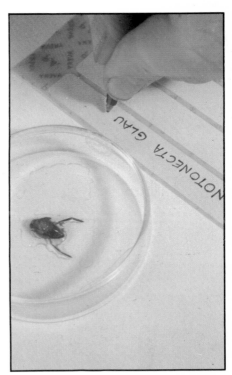

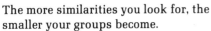
The more similarities you look for, the smaller your groups become.

The smallest basic group that biologists use is called a **species**. The word 'species' comes from a Latin word meaning 'kind' or 'sort'.

These animals have so many features in common that we recognize them as members of one species.

Every species has a Latin scientific name, but many have everyday names as well.

This individual has the everyday name of water boatman – but so do several other species. A more precise name is its Latin name – *Notonecta glauca*. The Latin name is more useful because there is a different one for each species and it means the same to everyone – whatever languages they understand.

Every living thing belongs to a species ...

Every animal belongs to a species
These animals have several features in common, but they can be divided into **four** different species.
Can you spot which is which?

Stoats *Mustela erminea*
are white underneath and have a
black tip to their tails.

Weasels *Mustela nivalis*
are white underneath and have
no black tip to their tails.

Polecats *Mustela putorius*
are brown underneath and have a
white patch behind their eyes.

Minks *Mustela vison*
are brown underneath and have
no white patch behind their eyes.

Every plant belongs to a species
These plants have several features in common, but they can be divided into **five** different species.
Can you spot which is which?.

Chives   *Allium schoenoprasum*
grow in clumps, and have narrow bulbs and narrow leaves.

Ramsons   *Allium ursinum*
grow in clumps, and have narrow bulbs and broad leaves.

Leeks   *Allium porrum*
grow singly, and have long cylindrical bulbs.

Onions   *Allium cepa*
grow singly, have round bulbs, and their flower stalks are swollen.

Garlic   *Allium sativum*
grow singly, have round bulbs, and their flower stalks are not swollen.

You don't have to look at these plants to recognize them – each species has its own smell. Try this for yourself at home.

Looks aren't everything
There are two different species here. Can you tell them apart?

Although they look almost identical, chiffchaffs and willow warblers belong to different species. The easiest way to tell them apart is to listen to their song. But once you know they are different, you can sometimes see small differences between them.

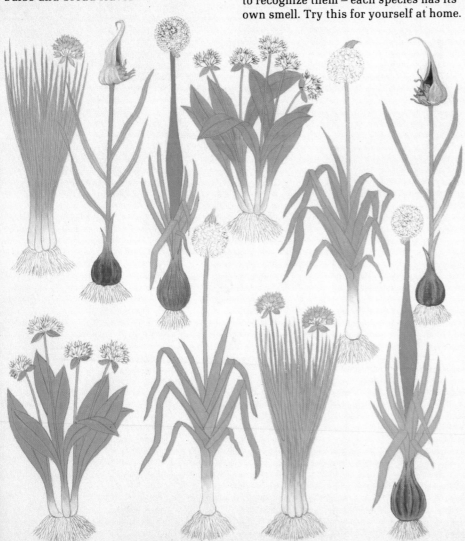

Chiffchaff
Phylloscopus collybita

Willow warbler
Phylloscopus trochilus

Different ... but still one species

No two people are exactly alike. But we are all members of the same species – *Homo sapiens*.

If you look at the oxeye daisies in a field, you will find that they have different numbers of petals.

There are many differences between individual oxeye daisies. But they all have so many features in common that we recognize them as a single species – *Leucanthemum vulgare.*

Other species are also variable. Later in the book, you can find out more about **variation** and the part it plays in natural selection.

How many species?

Tadpoles look very different from frogs, but they belong to the same species. They are different stages in the life of the same kind of animal – the common frog *Rana temporaria*.

When we sort living things into species, we must remember that they may look different at different stages of their lives.

This species doesn't always look like this . . .

This species doesn't always look like this . . .

it sometimes looks like this.

it sometimes looks like this.

A matter of opinion
Sometimes it is difficult to decide just how different individuals must be before we recognize them as two different species.

These two experts both worked on the difficult case of the dandelions.

In 1907, Alfred Rendle decided that these plants are **similar** enough for them all to belong to the same species.

In 1928, George Druce decided that the plants are **different** enough for each to belong to a different species.

Who is right? In some cases it is impossible to decide, and the number of species we choose to recognize is a matter of personal opinion.

Chapter 3

Species in nature

**Living things form their own groups.
Are these groups species?**

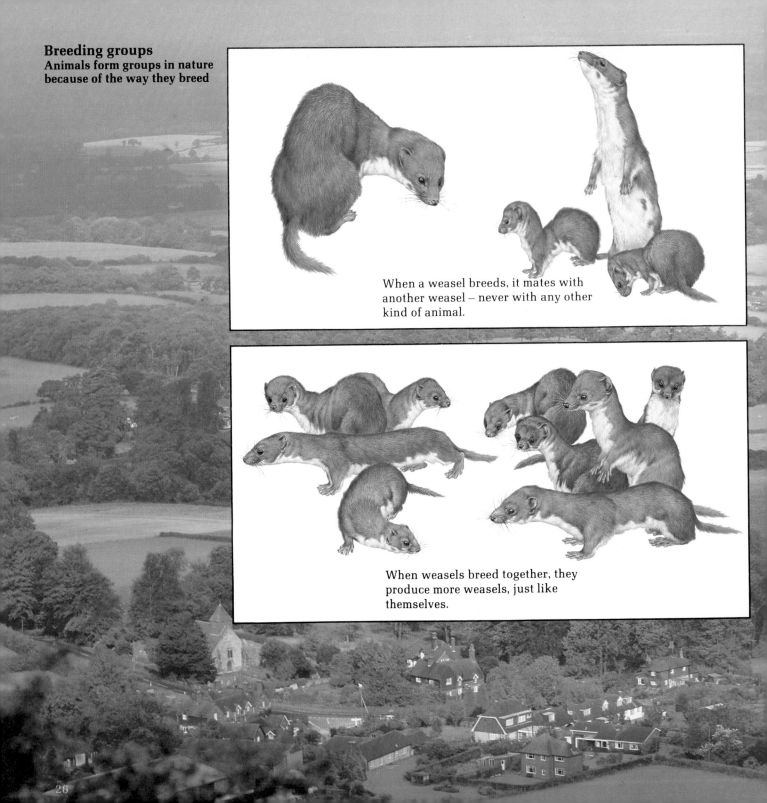

Breeding groups
Animals form groups in nature because of the way they breed

When a weasel breeds, it mates with another weasel – never with any other kind of animal.

When weasels breed together, they produce more weasels, just like themselves.

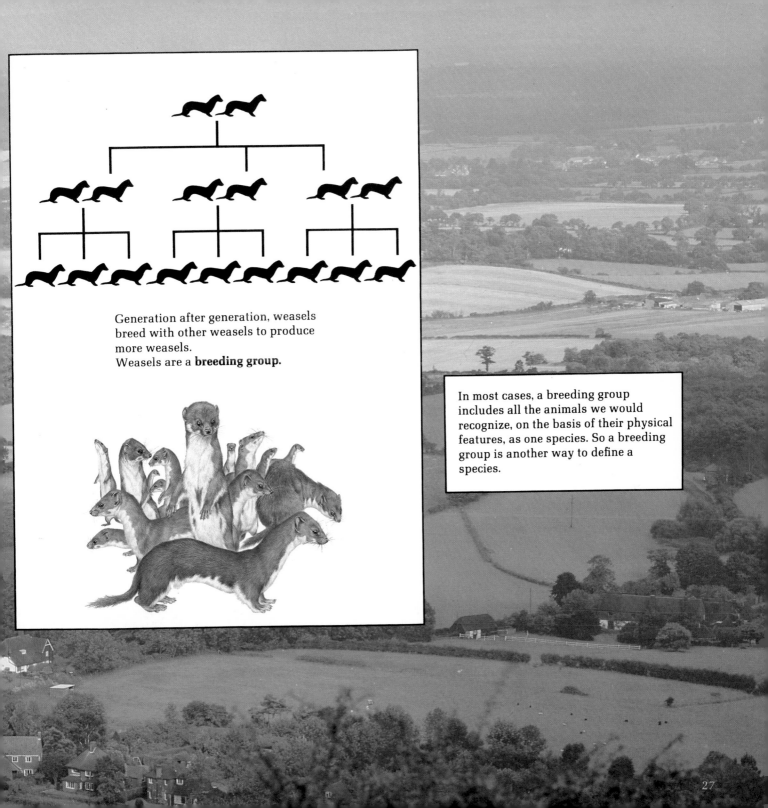

Generation after generation, weasels breed with other weasels to produce more weasels.
Weasels are a **breeding group.**

In most cases, a breeding group includes all the animals we would recognize, on the basis of their physical features, as one species. So a breeding group is another way to define a species.

Plants form groups in nature because of the way they breed
An oxeye daisy breeds with other oxeye daisies to produce more oxeye daisies. So oxeye daisies form a **breeding group**.

Other plants form breeding groups too. In most cases a breeding group includes all the plants we would recognize, on the basis of their physical features, as one species. So a breeding group is another way to define a species.

Barriers between breeding groups
A mule is a **hybrid** and can be produced only when a horse and a donkey breed together. Mules themselves cannot breed together.

Why is a mule like a fence?

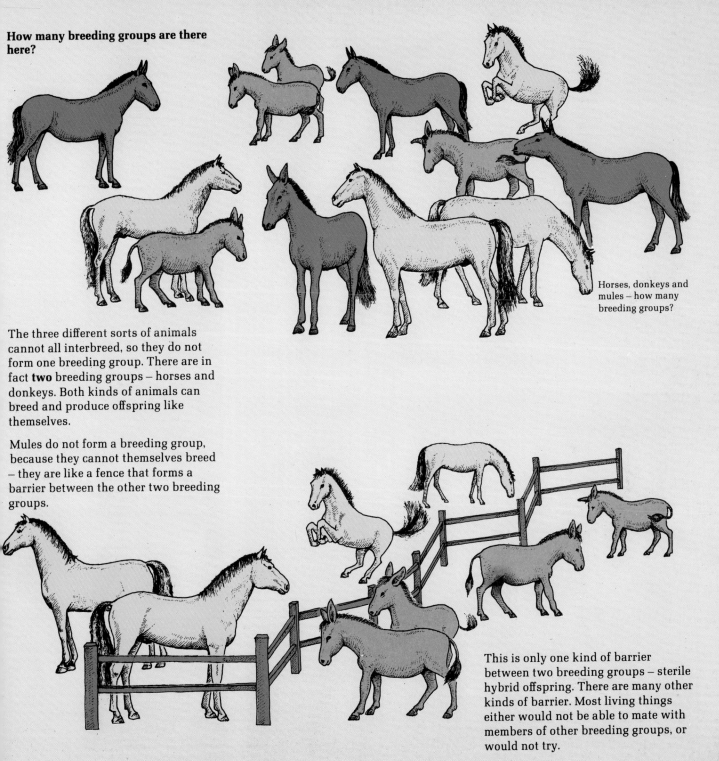

How many breeding groups are there here?

Horses, donkeys and mules – how many breeding groups?

The three different sorts of animals cannot all interbreed, so they do not form one breeding group. There are in fact **two** breeding groups – horses and donkeys. Both kinds of animals can breed and produce offspring like themselves.

Mules do not form a breeding group, because they cannot themselves breed – they are like a fence that forms a barrier between the other two breeding groups.

This is only one kind of barrier between two breeding groups – sterile hybrid offspring. There are many other kinds of barrier. Most living things either would not be able to mate with members of other breeding groups, or would not try.

29

What is a species?

In the last few pages, the word species has been used in two different ways ...
- to describe the groups we recognize,
- to describe the groups living things themselves recognize.

Are they the same?

In practice we can usually tell the difference between members of different breeding groups, and the word species means the same whichever way we use it. But sometimes the two definitions don't seem to correspond. When this happens, **breeding behaviour** is usually the best guide, as the following examples show.

All birds that look like this we recognize as king penguins, *Aptenodytes patagonica.*

One species that looks like two

There are two different sorts of foxgloves – smooth-stemmed and hairy-stemmed. We might think that they were two different species. But they interbreed, and are often found in the same population. So their differences are disregarded, and smooth and hairy foxgloves are considered to be one species.

Foxglove
Digitalis purpurea

Smooth-stemmed foxglove

King penguins form a breeding group in nature.

30

Two species that look like one

The opposite situation occurs when living things look similar, but don't interbreed. These fruit flies look so much alike that they seem to belong to one species. Yet they don't all mate with each other. There are two breeding groups here, so they are really two species.

If we look very closely at the flies, we can begin to find very slight differences between them. If we didn't know about their breeding behaviour, we might have put the two sorts of flies in the same species.

In cases like these, breeding behaviour is the best clue to the number of species.

Hairy-stemmed foxglove

Fruit fly
Drosophila pseudoobscura
about 30 times lifesize

Fruit fly
Drosophila persimilis
about 30 times lifesize

31

When there are no clues

Sometimes we don't know the breeding behaviour of plants and animals, so we cannot use this to help us sort them into species.

Here are three examples.

The solitary garlic A garlic plant does not produce seeds and never breeds with other plants. Instead, new plants grow from parts of its flower head or pieces of its bulb (cloves). So a garlic plant does not belong to a breeding group. But all garlic plants have so many features in common that we recognize them as one species.

The mysterious beetles These beetles come from the tropical rainforests of Sarawak, and have only recently been discovered. We can still only guess at their breeding behaviour, and have to rely on their appearance in order to sort them into species.

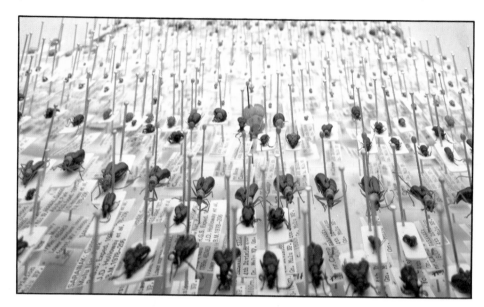

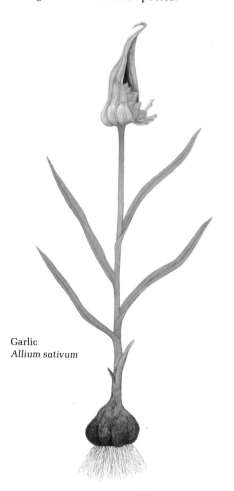

Garlic
Allium sativum

Ammonite
Asteroceras obtusum
about 180 million years old

The unknown past We can never have any definite information about the breeding behaviour of extinct living things, such as ammonites, that are known only from their fossil remains.

This is true for every fossil, even when there are very similar animals alive today, as in the case of the king crab.

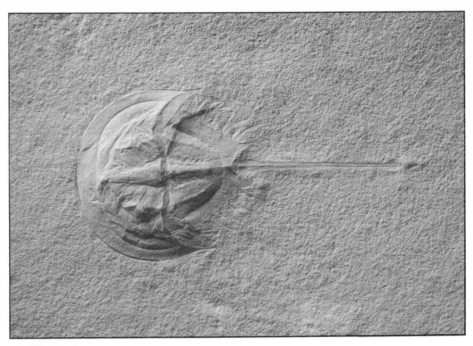

Fossil king crab
Mesolimulus sulcatus
about 140 million years old

King crab
Limulus species

So breeding behaviour cannot always be used to define species. But it is the best definition to use, because it depends entirely on the way living things themselves behave and is never a matter of personal opinion.

Darwin developed his theory from four different ideas about species. These ideas are introduced in the next four chapters.

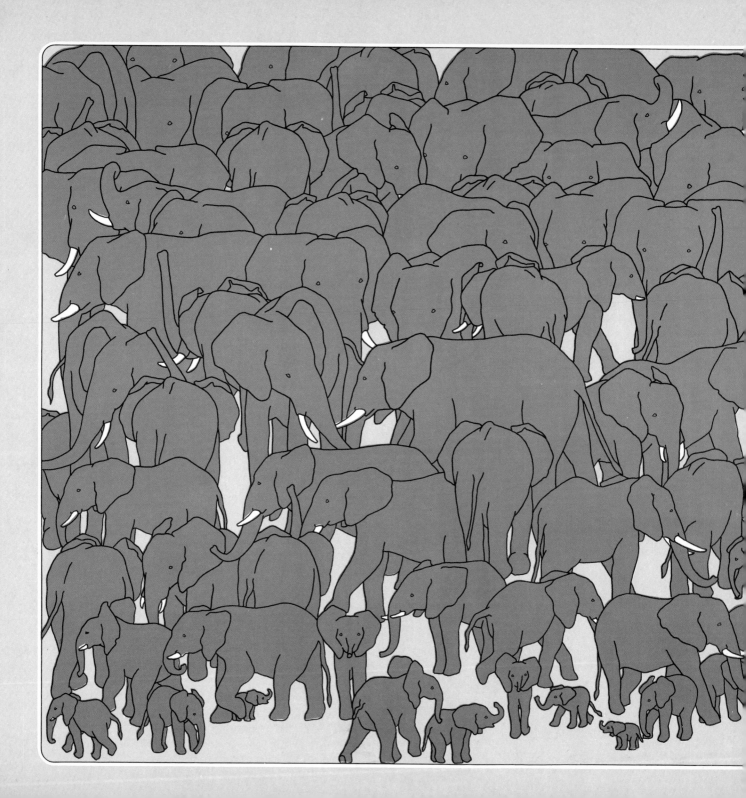

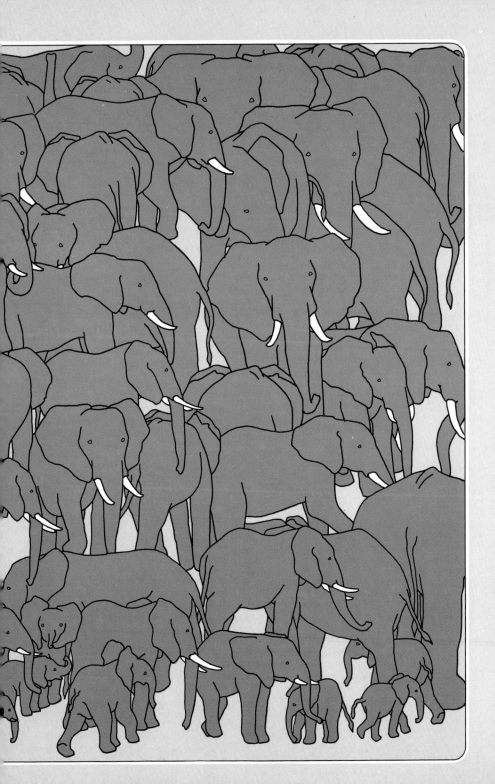

Chapter 4

More than enough offspring

Elephants breed more slowly than other animals. But Darwin worked out that, after 700 years, one pair of elephants could have 19 million descendants.

Possible numbers ...
Elephants unlimited!
A female elephant is able to have her first calf when she is about twelve years old ...

... And she could have another calf every two years. If the calves survived, they could all mate and have calves of their own ...

... and if these calves survived, they too could produce calves. So in every generation there would be many more breeding elephants ...

... and the number of elephants would increase more and more rapidly as time went on.

Just imagine how many elephants there could be!

More and more
Other species reproduce far more quickly than elephants.

After only 7 months, one pair of cockroaches could have **164 thousand million** descendants.

After 7 years, two poppy plants could have **820 thousand million trillion** descendants.

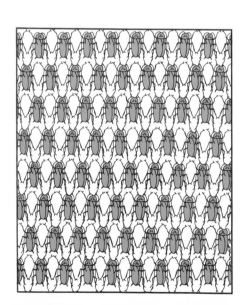

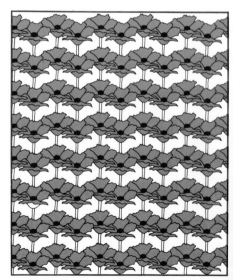

Actual numbers . . .

Compare the numbers of elephants in these pictures of the same area of natural and undisturbed African savannah.

100 years ago

You can see that the numbers are similar, despite the 100 year gap.

The same is true for most populations, wherever they live. Even though all species can produce more than enough offspring to replace themselves, numbers tend to remain the same if conditions do not change. This is because not all the offspring survive to reproduce.

In the next chapter you can find out some of the factors that affect survival.

Today

Chapter 5

The struggle to survive

How does the environment affect survival?

The environment of a rabbit
A rabbit, like all other living things, interacts with its environment.

What is the environment of a rabbit?

Other rabbits In order to produce offspring, a rabbit needs a mate. Rabbits are social animals, and build burrows near one another to form warrens.

Earth and air A rabbit lives on and under the ground. It breathes air and burrows in the soil for shelter.

Plants A rabbit is a herbivore, and depends on plants for food. Tall plants may also provide a hiding place.

Other animals Some animals eat the same food as a rabbit. Others, such as foxes and stoats, hunt rabbits to provide food for themselves. Fleas and other parasites may live in a rabbit's fur.

Weather Weather affects a rabbit in many different ways. When it is very wet, a rabbit tends to stay in its burrow. When it is fine, grass grows well, and provides plenty of food for a rabbit.

These are some of the many living and non-living things that make up the environment of a rabbit. They can all affect a rabbit's chances of survival.

41

The environment of an oak tree
An oak tree, like all other living things, interacts with its environment.

What living and non-living things make up the environment of an oak tree?

Weather Rain provides an oak tree with water, and sunshine provides the energy it needs to live and grow. Temperature affects a tree in many ways – for example, it influences the rate of growth and time of flowering.

Earth and air In order to live and grow, an oak tree needs light, air, water and nutrients. Its roots absorb water and nutrients, and anchor it into the soil.

Animals Many different animals feed on an oak tree, eating its leaves, its bark, its sap or its acorns. And many animals make their homes on or in the tree.

Plants Other plants growing near an oak tree may help to protect it. But, like an oak tree, they also need light and space and use water and nutrients.

All these different living and non-living things can affect an oak tree's chances of survival.

Not enough to go round

The resources of the environment are limited, and living things may **compete** for what they need.

In order to breed, a male rabbit needs a **territory** that includes females, burrows and a feeding area. But there are not enough territories to go round. So, in the breeding season, male rabbits compete with each other for territories.

Every year, thousands of tiny new oak trees are produced. Each one needs light and air, water and nutrients.

As the trees grow bigger, the environment cannot supply the needs of all of them. So there is competition for survival.

In the struggle to survive and reproduce, living things may compete with their own kind.

What are the effects of this competition?

The effects of competition
Winner takes all?

Red deer live together in herds. For most of the year, the stags (males) live apart from the females and young.

In late September, the stags move to the rutting (mating) grounds to find mates. Each of the older, stronger stags rounds up a herd of females. To keep his herd together, he must fight off all competitors.

The successful stags mate with most of the females in their herds. So these stags usually have many offspring.

The other stags have fewer offspring. Because they are unable to gather their own herds, they mate only with 'stray' females. The weakest stags may not mate at all.

One effect of competition is that some individuals have far more offspring than others.

45

Death from natural causes

Forty-five years ago, there were 60 000 young Scots pines growing amongst the older trees in this woodland. Of those 60 000 trees, only 1000 have survived to maturity.

What happened to all the others? Were they destroyed by some terrible disaster?

No, they died naturally ...

Young Scots pines are very small. As they grow older and taller, they need more and more light, space and nutrients. And, in the woodland, there is not enough for all of them. So the trees must compete with each other – and many die before they reach maturity.

One effect of competition is that only a few individuals survive to maturity.

Competition between species
Moles and blackbirds share a food resource – worms. Moles catch worms underground, and blackbirds catch them on the surface.

Imagine a situation in which moles and blackbirds depended entirely on worms for their food . . .

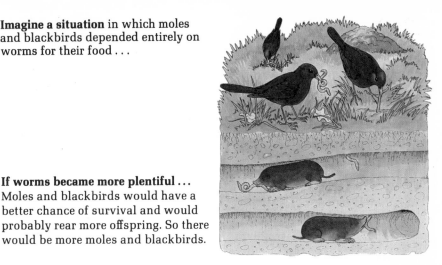

If worms became more plentiful . . .
Moles and blackbirds would have a better chance of survival and would probably rear more offspring. So there would be more moles and blackbirds.

If worms suddenly became scarce . . .
Competition for worms would become intense. So, many blackbirds and moles could die unless they could find alternative food.

If most of the moles died . . .
There would be more worms available for the blackbirds. So the blackbirds would have a better chance of survival, and would probably rear more offspring. And an increase in the number of blackbirds might affect other species . . .

The consequences
If ever different species need similar resources,
- changes in the resources,
- changes in the numbers of one of the species

may affect many other parts of the environment – and the survival of many other species.

Struggling to survive

Every year, a female lobster like this lays many thousands of eggs, which she carries attached to her body.

The eggs hatch and the larvae compete with each other for shelter and food. But, even if there is enough to go round, many do not survive.

What happens to them?

Lobsters are eaten by predators, including cod and octopuses.

Predators are part of the environment. They reduce the numbers of offspring that survive to reproduce.

Common octopus
Octopus vulgaris

European lobster
Homarus gammarus

48

Now we can begin to understand how the environment affects survival ...

All living things need food, space and suitable surroundings to live in. Without these, they cannot survive.

But the resources of the environment are limited. So living things **may** compete with one another for what they need.

Not all living things survive to produce young of their own. Sometimes survival is a matter of luck ... but some individuals have a better chance of producing young than others have.

Banded snail *Cepaea nemoralis*

Chapter 6

Some important differences

No matter what species you look at, you can always find differences between individuals.

Variation

All the members of a species are similar, but they are not identical.

For example, human beings are all one species, but there are so many differences between them that we have no difficulty in recognizing a face in a crowd.

If you look closely, you can find differences in other species too.

Lords-and-ladies
Arum maculatum

Banded snail
Cepaea nemoralis

Garden tiger moth
Arctia caja

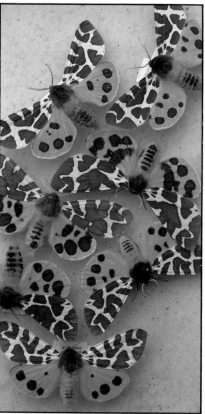

Lords-and-ladies are all one species, but individuals vary. Their leaves may be plain or spotted, and parts of their flowers may be purple or yellow.

These **banded snails** are all one species, but individuals vary. Their shells may be different colours, and may be plain or striped.

Garden tiger moths are all one species, but individuals vary. Their wings may have different markings.

Oxeye daisies are all one species, but individuals vary. They may be different sizes, and may have different numbers of petals.

Oxeye daisy
Leucanthemum vulgare

What causes variation?
Some variation is **inherited**. You can find out more about this in Chapter 7.

But some variation is caused by the **environment**.

Variation caused by the environment
Fading flamingos

In their natural environment, these flamingos have bright pink feathers. The colour is caused by a pigment, which the flamingos obtain from the microscopic algae that they eat. The flamingos cannot make the pigment themselves.

In captivity, the flamingos lose their bright pink colour unless they are given the correct diet.

**Variation caused by the
environment ... is not passed on**

Cuttings from the **same plant** ...

... develop very differently in **different
environments.** Cuttings that are grown
in the shade grow taller and more
spindly than those that are grown in
full sunlight.

But, if seeds from the two groups of cuttings are planted in the **same environment** . . .

. . . they grow into similar plants.

Variation caused by the environment cannot be passed on from one generation to another. This is true for all living things.

Chapter 7

A question of inheritance

It doesn't take long to work out which offspring belong to which parents. Living things resemble their parents.

A clear case of inheritance

Cuckoos lay their eggs in the nests of other birds. So young cuckoos are reared by birds of another species. But they grow up to resemble their real parents – in their appearance, their behaviour and in the way their bodies work.

Before she lays her own egg, a female cuckoo removes an egg from the host bird's nest.

Egg-laying

Egg-laying is controlled by hormones and occurs at different times in different species. Most birds lay their eggs in the morning, but cuckoos lay their eggs in the afternoon, when other birds are more likely to be away from the nest.

Behaviour

About ten hours after hatching, a young cuckoo starts to push the other eggs out of the nest. This behaviour increases the cuckoo's chances of survival by removing other young that would compete with it for food.

Living things **inherit** many characteristics from their parents. They inherit much of their appearance, their behaviour and the way their bodies work.

Like cow, like calf?

Inherited characteristics are controlled by genetic instructions – **genes** – which are carried on **chromosomes** in the nucleus of every cell. Most cells contain two matching sets of chromosomes – and hence two sets of genes. One set of genes is passed on from each parent during reproduction.

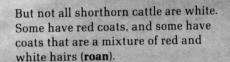

In shorthorn cattle, coat colour is inherited. White shorthorn parents always produce white calves.

But not all shorthorn cattle are white. Some have red coats, and some have coats that are a mixture of red and white hairs (**roan**).

What sort of calves would you expect these roan parents to produce?

Inheriting variation

The calves of roan parents are not all the same. Some are roan, like their parents, but some are red and some are white.

To understand why this variation occurs, we need to know more about the genes involved.

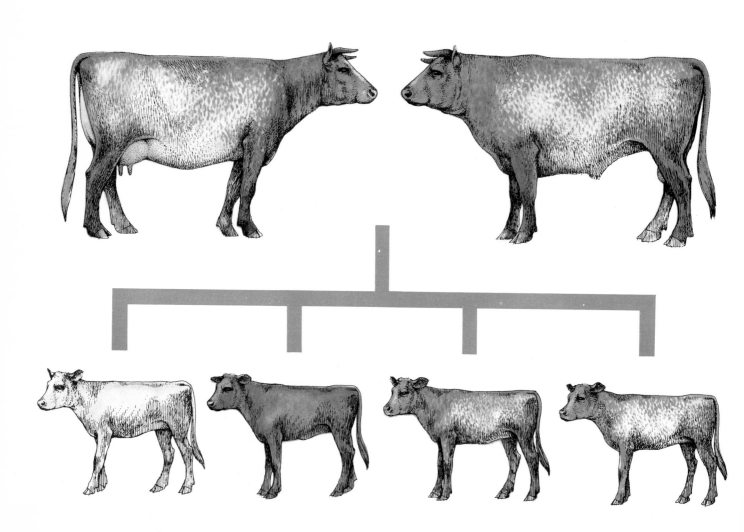

Genes and coat colour

Coat colour is controlled by a single pair of genes. There are two different genes for coat colour – one for red and one for white. A calf with a pair of 'white' genes is all white, and a calf with a 'red' pair is red. But a calf with a mixed pair of genes has a roan coat.

A calf inherits one coat-colour gene from each of its parents. These calves are different because they have each inherited a different combination of genes.

Genes and variation

This example shows the variation that a single pair of genes can produce. But all living things have thousands of pairs of genes controlling their appearance, their behaviour, and the way their bodies work. So the potential variation is enormous.

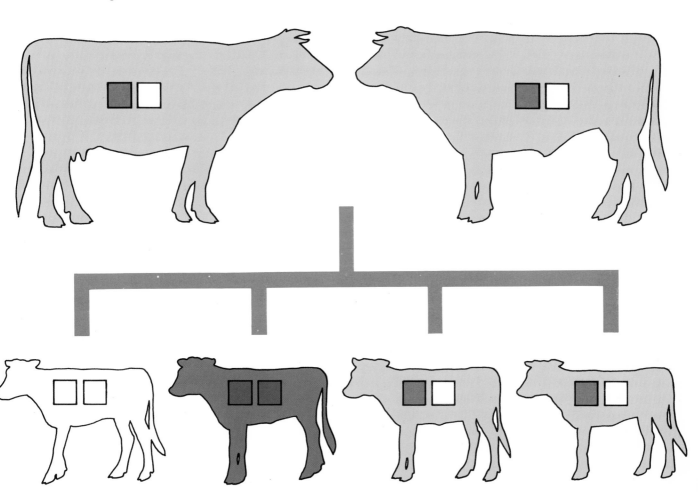

New variations

Sometimes a completely new variation occurs – a variation that can be inherited, but couldn't possibly be caused by a combination of the parents' genes. Here is one example.

The case of the short-legged sheep

1 In 1791 an unusual ram was born on Seth Wright's farm in New England ... it had short and crooked legs.

2 This set Seth thinking. If those short legs could be inherited, he would be able to breed a whole flock of short-legged sheep. Then he wouldn't need such high fences around his farm – and he could spend less on fence materials.

3 So Seth used the ram for breeding. Sure enough, two of its offspring had very short and crooked legs.

4 By breeding together these short-legged sheep, Seth eventually produced a whole flock of them – the Ancon breed.

A new gene—a new breed

Seth was lucky. The ram's short legs were caused by a changed gene – a **mutation**.

The ram passed on the mutation to some of its offspring – so they had short legs too.

By breeding together the short-legged sheep, Seth Wright eventually produced a whole flock of them – the Ancon breed.

He could never have produced a flock like the Ancons if he had just bred together the sheep with the shortest legs in his original flock.

Mutation and variation

Mutations are caused by a change in the chemistry of a gene or in the structure of a chromosome, and they are constantly occurring. In the case of the short-legged sheep, a mutation caused an obvious change. But many mutations have only slight effects or no effect at all. On the other hand, many mutations are harmful, or even lethal.

Haemophilia — a harmful mutation

Haemophilia is an inherited blood disorder caused by a mutation in a gene that affects the way blood clots. The blood of haemophilia sufferers clots very slowly, so they may bleed to death from minor injuries.

Many of Queen Victoria's descendants had haemophilia – they inherited the mutation from her.

This family tree shows the members of the Royal Family that were affected by haemophilia.

Trace the mutation through the generations.

Men who inherited the mutation suffered from haemophilia.

Women who inherited the mutation were **carriers**. They did not suffer from haemophilia, but were able to pass the mutation to their children. (Women suffer from haemophilia only if they inherit the mutation from both parents – and this did not happen amongst Queen Victoria's descendants.)

Eventually, haemophilia disappeared from the Royal Family.

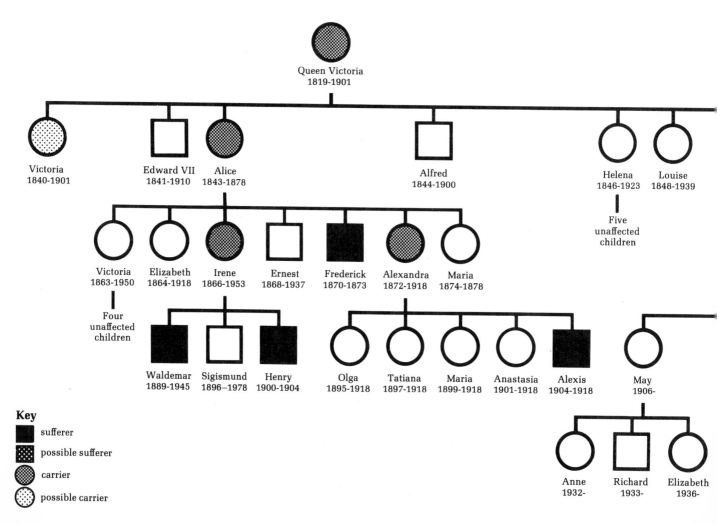

Key

■ sufferer
▨ possible sufferer
▦ carrier
⦿ possible carrier

Inheritance and variation

So now we can begin to understand some of the causes of the variation between individuals of the same species. Some variation is caused by the **environment**, and is not passed on from one generation to the next. Some variation is controlled by **genes**, and can be passed on from generation to generation. Much of the variation we see is a result of both the environment **and** genes.

The link between genes and inherited variation is rarely as simple as in the coat-colour example we described earlier. The study of inheritance – **genetics** – is an enormous subject and, for readers who would like to know more about it, there is a list of useful books on p 118.

Darwin himself knew nothing of the genetic mechanisms that control inheritance. All he knew was that some characteristics could be passed on from one generation to another. That knowledge enabled him to formulate his theory of natural selection . . .

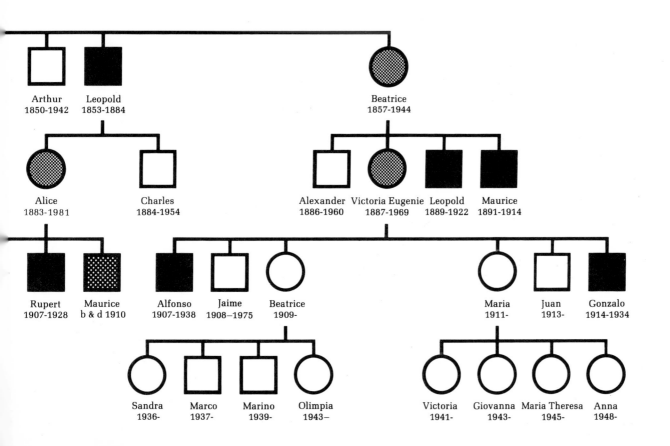

Arthur 1850-1942   Leopold 1853-1884   Beatrice 1857-1944

Alice 1883-1981   Charles 1884-1954   Alexander 1886-1960   Victoria Eugenie 1887-1969   Leopold 1889-1922   Maurice 1891-1914

Rupert 1907-1928   Maurice b & d 1910   Alfonso 1907-1938   Jaime 1908–1975   Beatrice 1909-   Maria 1911-   Juan 1913-   Gonzalo 1914-1934

Sandra 1936-   Marco 1937-   Marino 1939-   Olimpia 1943–   Victoria 1941-   Giovanna 1943-   Maria Theresa 1945-   Anna 1948-

1835
Oct 3d Island — The whole of this has the same
sterile dry appearance, studded with the
small craters, which are appendages to the
great Volcanic mounds — & from which in
very many places the Black Lava has flowed
the configuration of the streams being like
that so much mud. — I should think
it would be difficult to find in the intertropi-
cal latitudes, a piece of land 75 miles
long so entirely useless to man or the larger
animals. — From the evening of this
day to the 8th was most unpleasantly passed
in struggling to get about 50 miles to Windward
against a strong current. — At last we
8th reached James Island, the rendezvous of Mr.
Sulivan. — Myself, Mr Bynoe & three men
were landed with provisions there to wait
till the Ship returned from Watering. Chatham
Id. — We found on the Id. a party of
men sent by Mr Lawson from Charles Id. to Salt
fish & Tortoise meat (& procure oil from the
latter). — Near to our Birmaery place, there
was a miserable little Spring of water. —
We employed these men to bring us
sufficient for our daily consumption. —
We pitched our tents in a small valley
a little way from the Beach. — The little
Freshwater Bay was formed by two old craters: in this
Cove of ye island as in all the others, the mouths
Buccaniers from which the Lavas have flowed are
thickly studded over the country.
9th Taking with us a guide we proceeded

Chapter 8

Natural selection

What is natural selection and how does it work?

Darwin's theory

Darwin's theory of natural selection was based on the four ideas about species that have been introduced in the preceding chapters. It is easy to follow the steps in Darwin's argument if we apply these four ideas to a living population . . .

1 More than enough offspring

All species are capable of producing more than enough offspring to replace themselves.

One pair of **mice** can produce a litter of about six offspring as many as six times a year. Within six weeks, these offspring could produce litters of their own.

If all these mice survived and continued to breed, just imagine how many mice there could be . . .

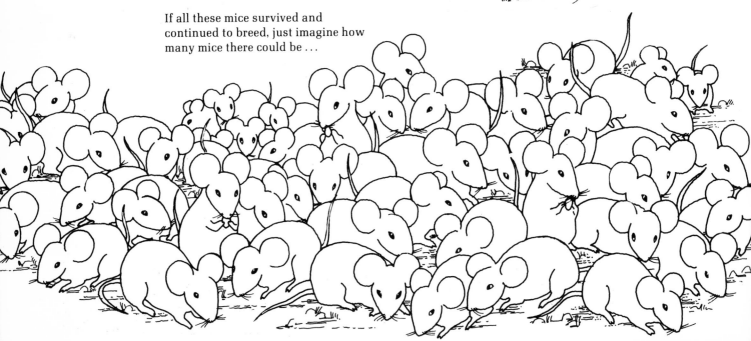

Why isn't the Earth covered with mice? Although a pair of mice can produce far more than enough offspring to replace themselves, the numbers in any population tend to remain more or less the same, because not all the offspring survive to reproduce.

2 The struggle to survive

The environment may affect an individual's chances of survival.

All living things interact with their environment. The environment provides food, space and suitable surroundings to live in, but it also includes competitors and predators. So, in any population, not all individuals survive to reproduce...

A mouse might be eaten by a predator.

Or it might not find a mate.

Or it might not be able to get enough food.

Or it might fail to find a nesting place.

Why do some individuals survive and not others? In some cases, it is a matter of chance which individuals die and which survive.

For example, a falling branch could land on any mouse that happened to be beneath it. In a situation like this, no mouse would have a better chance of surviving than any other.

If all mice were identical, survival would **always** be a matter of chance.

3 Some important differences

Because individuals are not all identical, some are more likely to survive than others.

No two mice are exactly alike, and some of the variations between them may affect their chances of survival.

Not all mice are the same colour – some are darker than others.

Mice are eaten by owls, which hunt for them by sight.

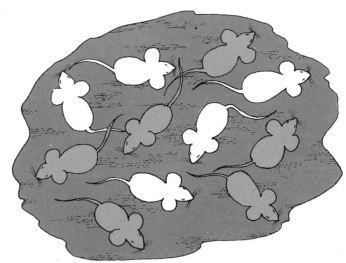

Against a dark background, the pale mice are easier to see, so they are the ones more likely to be eaten by owls. The dark mice are **better adapted** to this environment, and are more likely to survive.

The individuals that are best adapted to their environment are the ones most likely to survive and produce offspring.

4 A question of inheritance

Some characteristics are passed on to the next generation. Some of the variations between individuals are inherited. For example, mice inherit the colour of their coats.

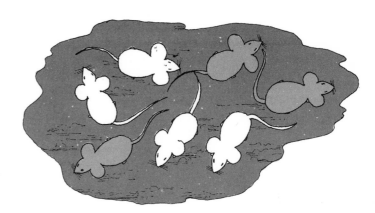

In an area of dark soil, dark mice are less likely to be seen by predators, and they have a better chance of surviving to reproduce. So their characteristics are the ones most likely to be passed on to the next generation.

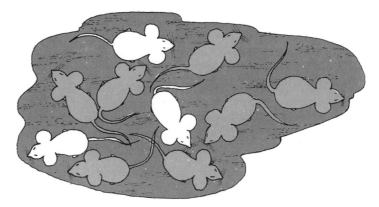

In the next generation, there will be a higher proportion of dark mice than before.

If conditions remain the same, the proportion of dark mice in the population will continue to increase.

Over many generations, the proportion of well-adapted individuals in a population is likely to increase. Darwin called this process **natural selection.**

Natural selection provides an explanation of how the characteristics of a population can change as individuals become better adapted to their environment. Natural selection can also have other effects on a population, as the following chapter shows.

Chapter 9

The effects of
natural selection

The theory of natural selection can be
used to explain a variety of situations
observed in nature.

Peppered moths — changing with the times

One effect of natural selection — the characteristics of a population can change.

Peppered moths are fairly common in Britain. All the specimens collected before 1848 looked like this . . .

Peppered moth –
pale form
Biston betularia

Then, in 1848, a dark moth of the same species was collected in Manchester.

Peppered moth –
dark form
Biston betularia

These two collections are typical of the peppered moth populations around Manchester in 1850 and 1900. The difference between the two populations is obvious.

How can this difference be explained by natural selection?

The offspring of dark moths are usually dark. So, if natural selection caused the change in the moth populations, the dark moths must have been better adapted to their environment than the pale moths were.
In what way?

The changing environment of the peppered moths

Several species of birds eat peppered moths, taking them from the tree trunks where they rest during the day.

During the nineteenth century, the moths' environment changed dramatically. Before the Industrial Revolution, most tree trunks had a mottled, grey appearance because they were encrusted with lichens.

Towards the end of the century, soot and smoke from factories killed most of the lichens and blackened the trees in many industrial areas.

Compare the two environments. In each situation, which moths are more likely to survive and leave offspring?

Manchester, 1850. There are six moths on this tree trunk. Which are better adapted to their environment – the pale moths or the dark ones?

50 years later. There are six moths on this blackened tree trunk. Which are better adapted to their environment – the pale moths or the dark ones?

Natural selection in action

Experiments have confirmed that birds find and eat more **dark** moths on pale, lichen-covered trunks. But they eat more **pale** moths on blackened trunks.

So in an industrial area, dark moths are more likely to survive and leave offspring. And, over a period of time, this would cause the proportion of dark moths in the population to increase.

This would help to explain the rapid change in the peppered moth species around industrial towns last century.

What's happening now?

In 1956, the Government passed the Clean Air Act, to reduce air pollution. Since then, the proportion of pale moths collected in industrial areas has gradually increased.
Why do you think this is?

Peppered moths show how natural selection can **change** the characteristics of a population.

Nuthatch eating a peppered moth

Manchester, 1980.

Changing with the times — other examples

Resistance to antibiotics

An antibiotic rarely kills all the bacteria it is intended to. The few bacteria that survive begin to reproduce. Because bacteria breed very quickly, the effects of natural selection can soon be seen. The population rapidly changes and, in a short time, it is made up of individuals that are not affected by the antibiotic. Such a population is said to be **resistant** to the antibiotic.

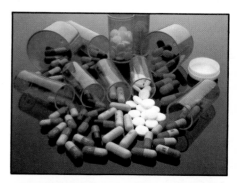

Resistance to insecticides

Just as bacteria can become resistant to antibiotics, insects can become resistant to insecticides.

We sometimes try to destroy harmful insects with insecticides. But an insecticide rarely kills all the insects it is intended to. Some survive and reproduce, and a resistant population builds up.

Newborn babies — a steady average

Another effect of natural selection — a well-adapted population stays that way.

Birthweight is a characteristic that is partly inherited.

Forty years ago, scientists made a study of almost 14 000 babies born in a London hospital. They found a link between the babies' birthweight and their chances of surviving the first 28 days after birth.

If human beings are a well-adapted species, we might expect most babies to have a good chance of survival.

Is this what the scientists found?

The results

What the babies weighed

*Numbers of babies

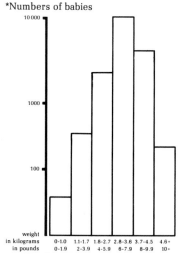

*plotted on a logarithmic scale

Some babies were heavier than others, but which was the **commonest** birthweight?

What happened to the babies

*Numbers of babies

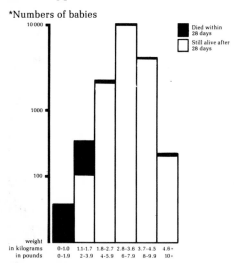

*plotted on a logarithmic scale

It looks as if some babies had a better chance of survival than others.

How can these results be explained by natural selection?

How birthweight affected survival

Not many **very small** babies were born, and few of them survived, usually because they were under-developed at birth.

Even fewer **very heavy** babies were born, and some of them died – mainly because of difficulties during birth.

Most babies were **medium-sized**, and had a very good chance of survival. Those that weighed 2.8–3.6 kilograms (6–8 pounds) had the best chance of all.

Natural selection in action

The study showed
- medium-sized babies were the commonest,
- medium-sized babies had the best chance of survival – they were the best adapted.

So medium-sized babies were the ones most likely to pass on their characteristics – including birthweight – to the next generation.

In other words, natural selection was maintaining a high proportion of the best-adapted birthweight – so a well-adapted population stayed that way.

White bark pines — a natural division

Another effect of natural selection — a population divides into two different types.

White bark pines live on the slopes of the Sierra Nevada Mountains in America. There are two types – **low bushes** and **tall trees**.

Their different appearances are mainly inherited. The two types can interbreed to produce hybrids of various shapes. But hybrids are very rare – most white bark pines are either low bushes or tall trees.

How can this be explained by natural selection?

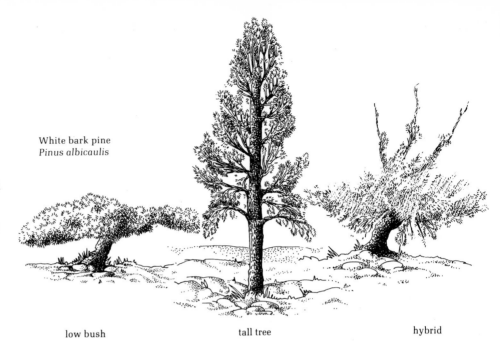

White bark pine
Pinus albicaulis

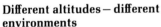

low bush   tall tree   hybrid

Different altitudes — different environments

The upper and lower slopes of the mountains provide completely different environments for the white bark pines.

The upper slopes are cold and windy. Strong winds dry out the plants and physically damage them. Frozen ground often prevents the roots from taking in water from the soil.

The lower slopes are mild and sheltered. Mild conditions allow the plants to get a regular supply of water from the soil. Trees shelter each other from the wind – but shade smaller trees from the light.

In each environment, which type of pine tree would be most likely to survive – the tall trees, the low bushes or the hybrids?

Natural selection in action

The low bushes are well adapted to the upper slopes and the tall trees are well adapted to the lower slopes, but the hybrids are not well adapted to either. So low bushes and tall trees have a good chance of surviving to pass on their characteristics to the next generation. But hybrids do not.

White bark pines show how natural selection can **divide** a population into two different types.

Natural division— another example

Many species have two sexes – males and females. The two sexes are often adapted to different roles, and have different characteristics. These differences may be maintained by natural selection.

For example, a male **golden pheasant** has brilliant plumage and performs a courtship display. Together these characteristics identify him to females as a suitable mate, and to other males as a competitor for mates. A male that is different may have less chance of finding a mate, and so his characteristics are less likely to be passed on to the next generation.

A female golden pheasant has dull plumage that conceals her from predators while she sits on her eggs. A female that looks different may have less chance of rearing offspring, so her characteristics are less likely to be passed on to the next generation.

Natural selection seems to be maintaining the differences between males and female in this species, and in many others.

Male golden pheasant *Chrysolophus pictus*

Male golden pheasants fighting

Female golden pheasant

Sickle-cell anaemia — a puzzling case

Another effect of natural selection — variation in a population is maintained.

Sickle-cell anaemia is an inherited blood disorder. Most sufferers are unlikely to survive to have children of their own.

As it is inherited, we might expect sickle-cell anaemia to have been eliminated by natural selection. But it is still common in some populations. How can this be explained?

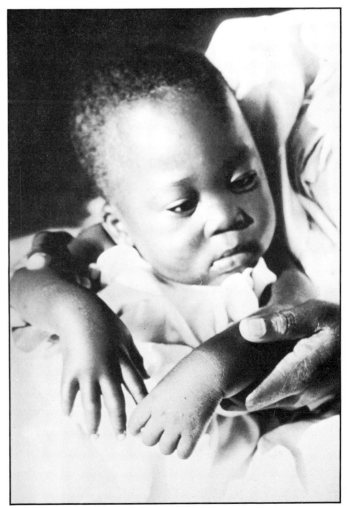

A child with sickle-cell anaemia

What is sickle-cell anaemia?

Sickle-cell anaemia affects the **red blood cells** that carry oxygen round our bodies.

Most people have normal red blood cells that look like this . . .

People who suffer from sickle-cell anaemia have red blood cells that look like this . . .

The cells have collapsed and become sickle-shaped, and cannot carry enough oxygen.

Some people have red blood cells that look like this . . .

Most of the cells look normal, but sometimes a few become sickle-shaped.

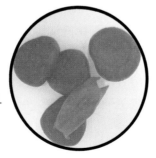

How is it inherited?

The characteristics that lead to sickle-cell anaemia are inherited through a single pair of genes. These genes control the production of **haemoglobin**, the oxygen-carrying substance in red blood cells. The normal form of haemoglobin is called **A**. There is also a slightly altered **S** haemoglobin that is caused by an abnormal gene.

Most people inherit a normal **A** gene from both parents, and we call these people **AA**. Their haemoglobin is normal.

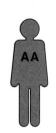

People who suffer from sickle-cell anaemia inherit an abnormal **S** gene from both parents. We call these people **SS**. All their haemoglobin is abnormal, and this is why their red blood cells become sickle-shaped.

Some people inherit an **A** gene from one parent and an **S** gene from the other. We call these people **AS**. Most of their haemoglobin is normal, but some of it is abnormal.

The puzzle

Most people with sickle-cell anaemia die before they have children. As a result, **S** genes are removed from the population. So you might expect sickle-cell anaemia to be very rare.

The clue

Sickle-cell anaemia is commonest in those parts of Africa where the most dangerous kind of **malaria** occurs.

Malaria is caused by a parasite that spends part of its life in red blood cells.

AS people are more resistant to malaria. (This is probably because the blood cells containing malaria parasites become sickle-shaped. So the parasites cannot complete their development.)

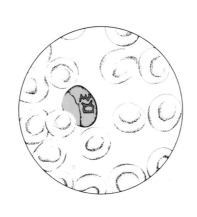

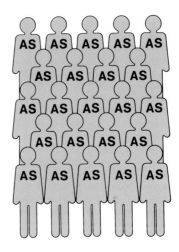

But in fact, sickle-cell anaemia is still quite common. Every year, thousands of **SS** babies die of sickle-cell anaemia and in some populations as many as 1 in 4 adults are **AS**.

How can this be explained?

The vital evidence

AA people have little resistance to malaria, and many of them die of the disease. As a result, **A** genes are removed from the population.

Because **AS** people are unlikely to die from malaria, their genes, **A** and **S**, are kept in the population.

Because they are resistant to malaria, in some areas **AS** babies are the ones most likely to survive to have children. If two **AS** people marry, 1 in 4 of their children are likely to be **SS** and to suffer from sickle-cell anaemia.

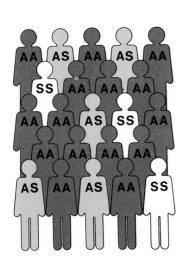

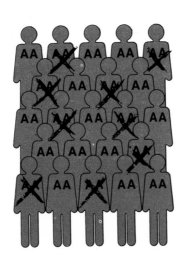

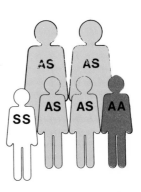

Summing up—a question of balance

Now we can explain why sickle-cell anaemia remains common in malarial areas.

One effect of natural selection is that **A** and **S** genes are **removed** from the population by malaria and sickle-cell anaemia.

Another effect of natural selection is that **A** and **S** genes are **kept** in the population because **AS** people are more resistant to malaria.

There is a balance between these two effects of natural selection. The proportion of **A** and **S** genes in the population depends on how common malaria is.

Sickle-cell anaemia shows how natural selection can maintain genetic variation in a population.

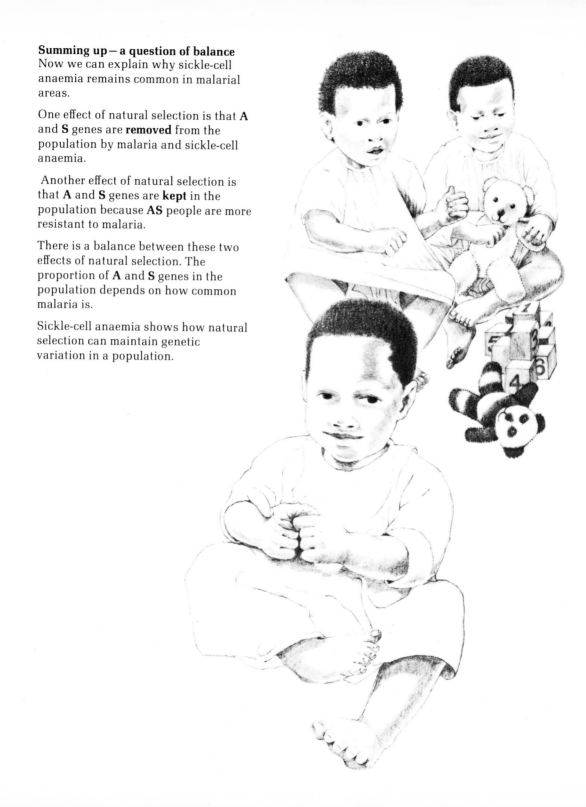

Pipe-vine and tiger swallowtails – a case of mistaken identity

Sometimes natural selection is not a constant force.

Pipe-vine swallowtail butterflies live in North America. They taste so unpleasant that birds soon learn not to eat them.

Tiger swallowtail butterflies, which live in the same area, are perfectly good food and much preferred by birds. There are two kinds of female – some are stripey, like the males, but others are dark and look like pipe-vine swallowtails. The dark females are called **mimics**.

Which kind of female is more likely to survive?

The mimics are the ones more likely to survive, because they are easily mistaken for the distasteful pipe-vine swallowtails. So, birds that have learnt to avoid pipe-vine swallowtails avoid the mimics as well.

If natural selection is acting here, we might expect the mimics to eventually replace the stripey females. But this has not happened – probably because, if there are a lot of mimics, birds may not learn to associate dark butterflies with unpleasant taste. So the chances of the mimic swallowtails surviving to reproduce depend on how many of them there are.

In the case of the tiger swallowtails, natural selection is not a constant force – it depends on the numbers of the different types of individual in the population.

Pipe-vine swallowtail
Battus philenor
female

Tiger swallowtail
Papilio glaucus
female

Tiger swallowtail
mimic female

Pipe-vine swallowtail
male

Tiger swallowtail
male

Mocker swallowtail butterflies
Natural selection can act in several different ways at once.

Mocker swallowtail butterflies live in Africa. All the males look the same, but there are at least four different types of female. One type is black and yellow, like the males, but the others are mimics, each resembling a different species of butterfly.

Similar males

The black-and-yellow pattern of a male identifies him to females as a suitable mate. A male with a very different pattern would probably have less chance of mating and passing on his characteristics to the next generation. So natural selection tends to keep the pattern more or less the same.

Different females

The mimic females each resemble a different species of distasteful butterflies called **danaids**. Mocker swallowtails themselves are not distasteful. Because birds mistake them for the distasteful danaid butterflies, the mimic females are more likely to survive than females with other patterns. So natural selection maintains the three distinct types of females.

Mocker swallowtail
Papilio dardanus
female

Mocker swallowtail
male

Mocker swallowtail
mimic female

Mocker swallowtail
male

*Danaus chrysippus***

Different numbers

If the mimics were as numerous as the danaid butterflies, birds would be unlikely to associate the wing patterns with unpleasant taste. So, for any one kind of mimic female, the more there are, the less their mimicry protects them.

By imitating several different species, mocker swallowtails maintain a larger protected population. The proportions of the different kinds of female are controlled by natural selection.

A further change

In Ethiopia, the mimic female mocker swallowtails have 'tails' on their wings – like those of the males, but smaller. The distasteful butterflies that they mimic have no tails at all. Could natural selection be in the process of changing wing shape in this population?

Mocker swallowtail
mimic female

Mocker swallowtail
mimic female

Mocker swallowtail
mimic female
from Ethiopia

Amauris echeria ∗

Amauris niavius ∗

∗ Danaid butterflies

Banded snails
– a continuing puzzle

An assortment of shells

Banded snails are widespread in open country and woodland in Britain and Europe. Their shells are extremely variable – they may be yellow, brown or pink, and some have dark brown bands that spiral round the shell (as you can see on pages 51 and 52). These shell patterns are inherited, and are not caused by the environment. A single population of snails may contain many different shell patterns.

In some areas, the proportions of the different kinds of shells vary according to the habitat. Around Oxford, for example, there are more yellow shells in open grassland and more pink or brown shells among the dead leaves in woodlands. And the proportion of plain snails is greater in a uniform habitat, such as a carpet of dead leaves, than in a mixed habitat, such as rough grassland or a hedgerow.

The variation in the proportions of the shell patterns can be explained partly by natural selection.

The thrushes' selection

Snails are eaten by birds, especially thrushes. First the birds have to remove the snails from their shells – they carry them to a suitable stone and drop or tap them until the shells break. If we compare the patterns on the broken shells with those of the population as a whole, we find that thrushes usually pick out the snails that are most easily seen against their background.

The snails that are least easily seen have the best chance of surviving to reproduce. Natural selection favours the best-adapted shell patterns in each habitat, and so populations in similar habitats tend to resemble one another. But differences do occur and each population has its own assortment of shells. How can this be explained?

Banded snail
Cepaea nemoralis

Which snails are easiest to see against each of the different backgrounds? But which are most likely to survive?

Other influences

Other factors, such as climate, also influence the survival of banded snails. In some populations, snails with dark shells are better adapted to cold conditions and pale snails are better adapted to hot conditions. A simple explanation would be that the different colours absorb heat differently. But not all populations are affected by the climate in this way.

In some places, we cannot see any reason for the variation in the shells, particularly in their banding patterns. And recent studies have shown that the variety of shells in one population may be quite different from that in other populations living nearby. The reasons for this are not understood – perhaps we don't understand the habitat well enough.

Natural selection or not?

There are still many problems to be solved before we can fully explain all the variation in banded snails. To understand the part played by natural selection, we need to find out how well each shell pattern is adapted to each different environment. This is an almost impossible task – partly because there are so many genes involved, and partly because it is extremely difficult, in a living population, to test all the relevant factors.

It is therefore difficult or impossible to predict the effects of natural selection in nature. The results depend so much on the context, because each locality is unique and each population has a different range of shell patterns. At the moment all we can say is that natural selection is one of the many factors that must be taken into account in the story of the banded snail.

The effects of natural selection

Summary
As a result of natural selection, living populations become better adapted to their environment.

If the environment changes, the characteristics of the population will change as the population becomes better adapted to the new environment. This is what happened to the peppered moths around Manchester last century.

But natural selection does not always lead to change. The example of babies' birthweight shows how, in an unchanging environment, natural selection ensures that a well-adapted population stays that way.

Where the environment varies, natural selection may divide a population into two or more distinct types – each adapted to a particular part of the environment. The white bark pines are a good example of this.

In all these cases, the effects of natural selection are easy to see. But, as the banded snails show, in most living populations so many different factors have to be taken into account that it is much harder to spot the effects of natural selection.

The examples described in this chapter illustrate natural selection in action amongst living species. But what part does it play in the formation of new species?

Chapter 10

How are new species formed?

These macaws do not interbreed, so
they are all different species.
How are species formed?
What part does natural selection play
in the process?

91

Forming a new species

We can never go back in time, so we can never be certain how living species are formed. But observations of living species can provide many clues.

Over the years, people have suggested a variety of mechanisms for the origin of species. Darwin himself suggested that each new species is formed over many thousands of years, by a process involving natural selection. But natural selection is only part of the story. More recently, scientists have recognized three distinct stages in the formation of new species ...

Stage 1
A barrier to breeding
One species may be split into two separate populations by some sort of **barrier** which prevents the two populations from breeding together. The barrier is often a geographic barrier, such as a mountain range.

Stage 2
Becoming different

Each population becomes adapted to its local environment – usually under the influence of natural selection. Over many generations, the two populations may gradually become **different**.

Stage 3
Two different species

The two populations may eventually become so different that they can no longer interbreed. The two populations have become **two different species**.

Let's look more closely at the three stages in the formation of new species.

Stage 1

A barrier to breeding

The first stage in the formation of a new species is when one species is split into two separate populations by some sort of barrier. The two populations **could** still interbreed, but the barrier prevents them from doing so. The barrier that prevents the two populations from interbreeding is often a **geographic barrier**.

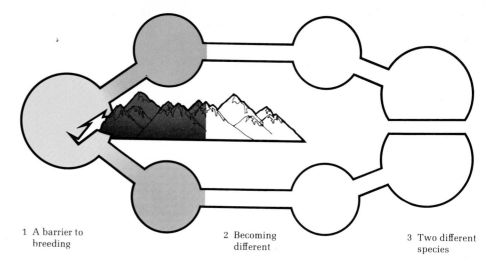

1 A barrier to breeding

2 Becoming different

3 Two different species

Geographic barriers

A geographic barrier is an area where a species cannot live. It could be ...

... mountains

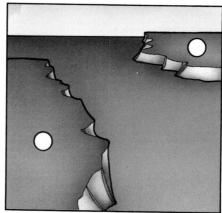

... sea

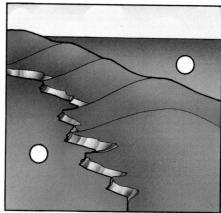

... land

... a river

... a desert

... ice

A barrier to breeding

Sea

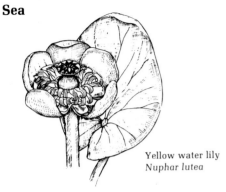

Yellow water lily
Nuphar lutea

Yellow water lilies are common in ponds, ditches and slow-moving rivers throughout most of Europe. They are all one species, but the English Channel prevents the water lilies in England from interbreeding with those on the Continent.

Fossil pollen grains show that yellow water lilies were growing in England and many parts of Europe before the English Channel was formed. So at that time the water lilies would all have been able to interbreed. But now the English Channel forms a **geographic barrier** between the water lilies in England and those on the Continent.

Pollen grain of a yellow water lily about 1200 times lifesize

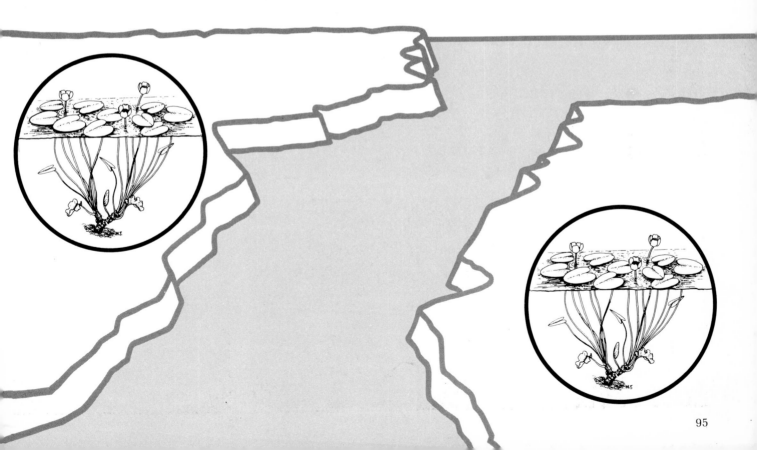

Stage 2
Becoming different

Now they are separate, the two populations may be in slightly different environments.

Over many generations, they become adapted to these environments under the influence of natural selection. So the two populations gradually become different.

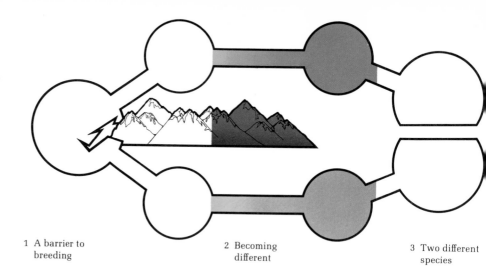

1 A barrier to breeding

2 Becoming different

3 Two different species

Different environments

Even in a small area, there may be differences in the environment, and the two populations may become adapted to...

...different rainfall

...different backgrounds

...different temperatures

...different altitudes

...different terrain

...different depths

Becoming different
The effect of altitude

Sticky potentillas are shrubby plants that grow in California.

Although they are all one species, they are not all the same. High on the mountains they are short and stunted, but on the coast they grow tall and bushy.

The different types are adapted to different environments. The short stunted plants are well adapted to the cold and windy mountain environment, and the tall plants are well adapted to the more sheltered coastal environment.

If a coastal plant is planted on the mountains, it soon dies.

Many other plant species have mountain and lowland forms. Examples include the white bark pines described in the last chapter.

Stage 3

Two different species

The two populations may eventually become so different that they can no longer interbreed – even when the barrier is no longer there.

If the two populations become two separate breeding groups, they are now **two different species**.

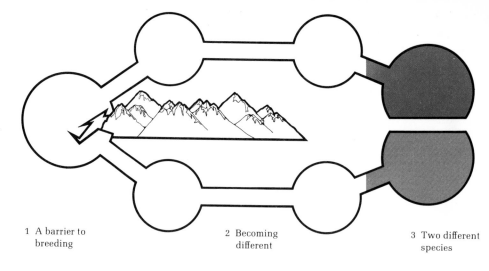

1 A barrier to breeding

2 Becoming different

3 Two different species

What prevents interbreeding?

Two populations may be unable to interbreed if . . .

. . . they live in different habitats

. . . they look different

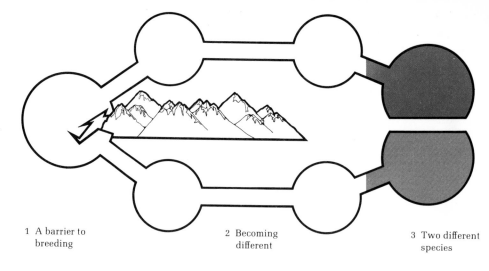

. . . they behave differently

. . . they breed at different times of the year

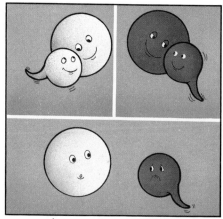

. . . their sex cells are incompatible

. . . their offspring do not survive, or are sterile

What prevents interbreeding?

Two species — two mating calls

These grasshoppers look almost identical, but they belong to two different species. The most obvious difference between them is the mating calls that the males make to attract the females.

Female grasshoppers recognize males of their own species by their mating call. Normally, they do not mate with males of other species, because their mating calls are different.

But, in the laboratory, a female grasshopper can be 'tricked' by a tape recording of the mating call of her own species. When she hears this call, she will mate with a male of the other species which has been placed in her cage.

The hybrid offspring that result from this mating are healthy and fertile. So it seems that it is only the different mating calls that prevent the two species from interbreeding.

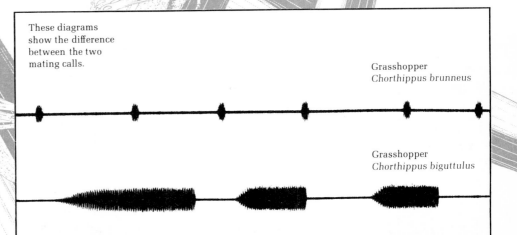

These diagrams show the difference between the two mating calls.

Grasshopper
Chorthippus brunneus

Grasshopper
Chorthippus biguttulus

Stage 3

Some different endings

The first part of the chapter showed that, when a species is divided into two, the separate populations may eventually become so different that they can no longer interbreed. And, as described in Chapter 3, two populations that cannot interbreed are two different species.

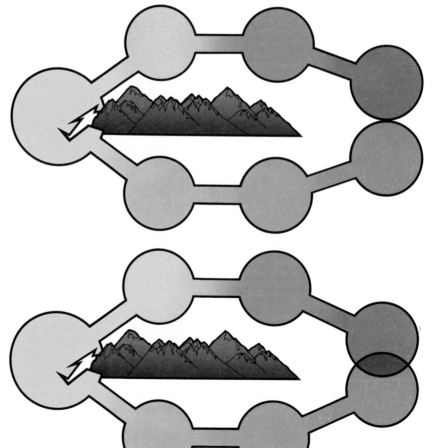

But the ending can be different ...

Even though the two populations have become so different that they **look** like different species ...

... they may still be able to interbreed.

What happens next depends on which individuals are the best adapted – members of the two parent populations, or their hybrid offspring.

What will the ending be . . .

... if the hybrids are better adapted
than the two parent populations?

... if the hybrids are not as well
adapted as the two parent populations?

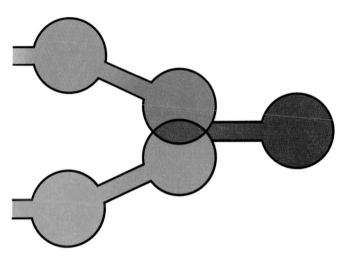

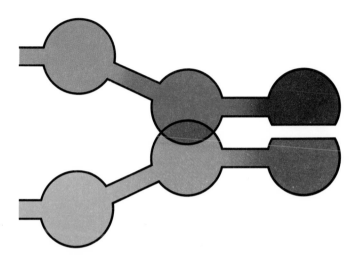

A **new hybrid species** may be formed.
The hybrids are better adapted than
the parent populations – in other
words, they are the ones most likely
to survive. If the hybrids are fertile,
they may eventually replace the
parent populations, and could be
recognized as a new species.

Two species may be formed. The
parent populations are better adapted
than the hybrids – in other words,
they are the ones more likely to
survive. So eventually the populations
may become two different species.

You can see examples of these
different endings on the next two
pages.

One species or two?

Bishop pines and Monterey pines grow on the Monterey peninsula in California. They look quite different, but they are capable of interbreeding to produce hybrids. The hybrids are not as well adapted as the parent species, and do not survive.

Bishop pine
Pinus muricata

Monterey pine
Pinus radiata

The two sorts of pine trees breed at different times of the year. Monterey pines shed their pollen early in the year, and Bishop pines shed their pollen later.

So the two trees do not normally interbreed, and are considered to be two different species.

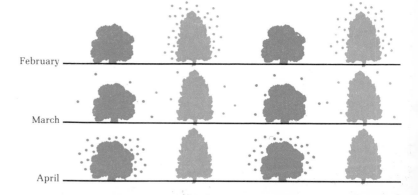

February

March

April

A different ending

In England there are two species of hawthorn, which can easily be distinguished by their leaves. Common hawthorns live in open places, and midland hawthorns live in dense woods.

A new habitat, a new species?

As woodlands are cleared for agriculture, the habitat of the midland hawthorns is gradually being destroyed, but common hawthorns can grow in the clearings. So the two species now grow near each other and can, in fact, interbreed. The hybrids are fertile, and can be distinguished by their intermediate leaves.

The hybrids are better adapted to the new intermediate habitats than the other hawthorns are, and they are becoming more and more widespread. One day the hybrids may entirely replace the midland hawthorns. Would they be recognized as a different species?

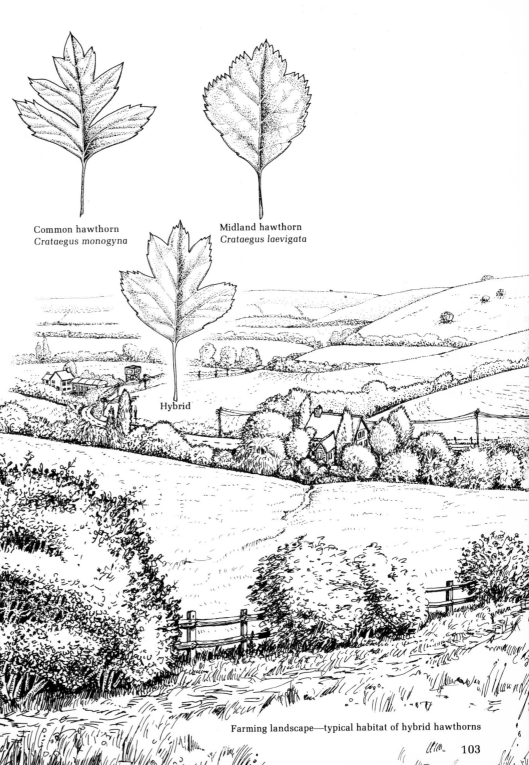

Common hawthorn
Crataegus monogyna

Midland hawthorn
Crataegus laevigata

Hybrid

Farming landscape—typical habitat of hybrid hawthorns

103

New species are formed in other ways

A species can be split into two populations even when all the individuals live in the same area. For example, the two populations might eat different foods, or have different breeding seasons. The two populations are then separated by an **ecological barrier,** and cannot interbreed.

An ecological barrier
Different breeding seasons

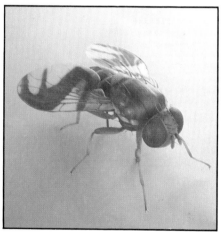

Fruit fly
Rhagoletis pomonella
about 6 times lifesize
(photographed from a model)

This species of fruit fly lives in North America. It is divided into two populations that live on different trees in the same area. Some live on hawthorn trees and some live on apple trees. The two populations do not interbreed.

How could this have come about?

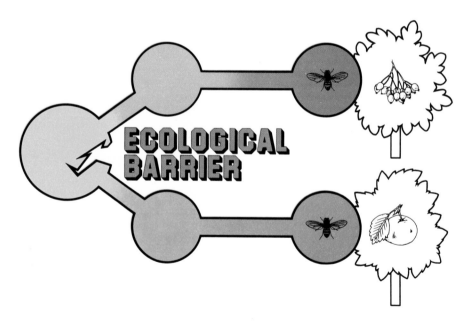

Once there were only 'hawthorn' flies
Over a hundred years ago, all the flies lived on hawthorn bushes. Part of their life cycle is like this ...

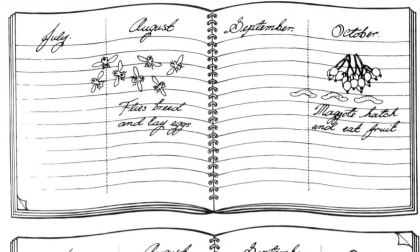

Now there are 'apple' flies too A hundred years ago, apple trees were introduced into the area. Now some fruit flies live on apple trees. Their life cycle is like this ...

Why do 'apple' flies breed earlier than 'hawthorn' flies? Apples tend to ripen earlier than hawthorn fruits. Not all the flies hatch at the same time and, when apple trees were introduced into the area, perhaps some of the flies hatched early enough to feed on apples rather than hawthorn fruits. Over a period of time, natural selection has divided the fruit flies into two quite separate populations, breeding at different times.

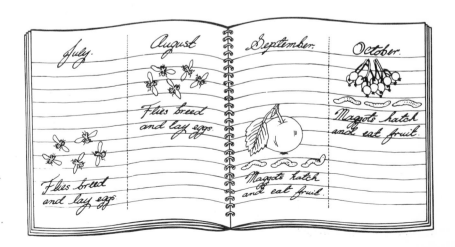

Are there now two species?
'Apple' flies and 'hawthorn' flies look the same and they **can** still interbreed. But they do not, because they breed at different times. Eventually the two kinds of fruit fly may become **two different species.**

Instant species

We have seen that it can take a very long time for one species to become two. But sometimes the hybrid offspring of two different species can become a new species almost at once, because of changes in their chromosomes.

Usually hybrids are sterile, because the sets of chromosomes they get from each parent don't match and cannot be passed on. Some hybrids can become fertile by **polyploidy** – a process that results in offspring with more sets of chromosomes than their parents. Polyploidy is a complicated process, and readers who would like to know more about it should refer to the booklist on p.118.

Polyploidy is rare in animals, but common in plants. **Swedes**, for example, are a polyploid species formed from a hybrid between a type of cabbage and a type of turnip. (Because no one knows exactly what these early vegetables were like, they are represented here by their modern relatives.)

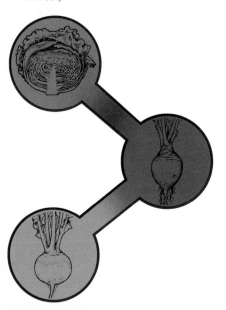

All these plant species are polyploids.

Sun spurge
Euphorbia helioscopia

Cocksfoot grass
Dactylis glomerata

Townsend's cordgrass
Spartina × townsendii

Cloudberry
Rubus chamaemorus

Dog rose
Rosa canina

Annual meadowgrass
Poa annua

Common hempnettle
Galeopsis tetrahit

Wild pear
Pyrus communis

Common whitebeam
Sorbus aria

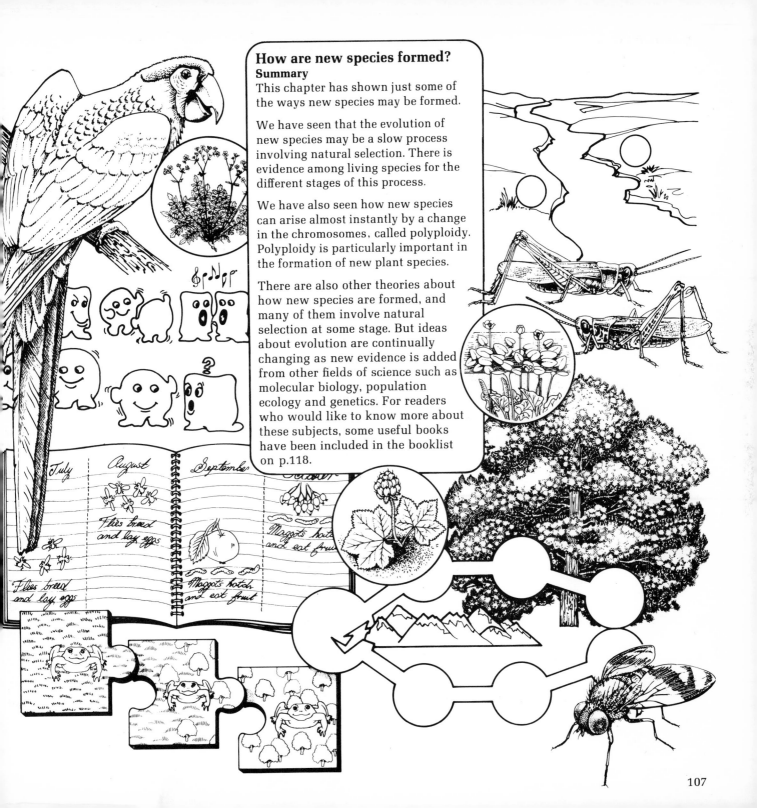

How are new species formed?

Summary

This chapter has shown just some of the ways new species may be formed.

We have seen that the evolution of new species may be a slow process involving natural selection. There is evidence among living species for the different stages of this process.

We have also seen how new species can arise almost instantly by a change in the chromosomes, called polyploidy. Polyploidy is particularly important in the formation of new plant species.

There are also other theories about how new species are formed, and many of them involve natural selection at some stage. But ideas about evolution are continually changing as new evidence is added from other fields of science such as molecular biology, population ecology and genetics. For readers who would like to know more about these subjects, some useful books have been included in the booklist on p.118.

July   August   September

Flies breed
and lay eggs

Flies breed
and lay eggs

Maggots hatch
and eat fruit

Maggots hatch
and eat fruit

A view of the Galapagos Islands, looking towards San Salvador

Chapter 11

Origin of species

Is there any evidence in the natural
world that natural selection can lead to
the evolution of new species?

The Galapagos finches

On the Galapagos Islands there are 13 species of **finches** that are found nowhere else in the world. They were first studied by Charles Darwin when he visited the islands in 1835 during the voyage of HMS *Beagle*.

The Galapagos finches are drab little birds, but Darwin noticed that each species had a different shape or size of beak. His observations of the finches were to become one of the strongest arguments for the role of natural selection in the origin of species – and today the finches are still considered to be a textbook example of evolution.

Mangrove finch
Cactospiza heliobates
Lives on only two islands in the west.

Woodpecker finch
Cactospiza pallida
Lives on the central islands.

Vegetarian finch
Platyspiza crassirostris
Lives on many of the islands.

Warbler finch
Certhidea olivacea
Lives on all the islands.

Small tree finch
Camarhynchus parvulus
Lives on nearly all the islands.

Large tree finch
Camarhynchus psittacula
Lives on many of the islands.

Large tree finch
Camarhynchus pauper
Lives on only one island in the south.

Cactus finch
Geospiza scandens
Lives on all except the most northerly islands.

Large cactus finch
Geospiza conirostris
Lives on only three islands.

Large ground finch
Geospiza magnirostris
Lives on nearly all the islands.

Medium ground finch
Geospiza fortis
Lives on nearly all the islands.

Small ground finch
Geospiza fuliginosa
Lives on nearly all the islands.

Sharp-beaked finch
Geospiza difficilis
Lives only on the more northerly islands.

Colour illustration by John Gould, 1841.

A closer look at the Galapagos Islands

The Galapagos Islands are in the Pacific Ocean about 1000 kilometres from the South American mainland. They are the remains of volcanoes that rose up from the sea only a few million years ago.

The few species of plants and animals that live on the Galapagos Islands must be descended from ancestors that originally came from the mainland. Darwin noticed that they were very different from the mainland species, and thought that this could be explained by evolution.

The Galapagos Islands

Marchena

Scale in kilometres

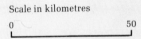

San Salvador

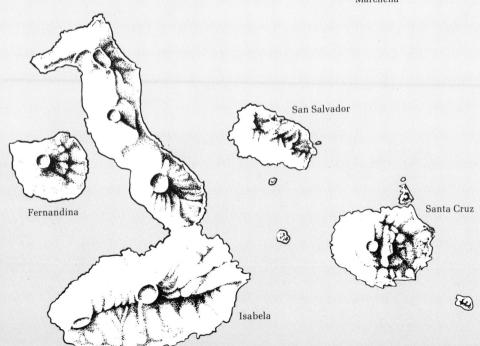

Fernandina

Santa Cruz

San Cristobal

Isabela

The right tool for the job

Darwin noticed that there were several species of finches on the Galapagos Islands and that each had a different shape or size of beak. The differences can be explained by **natural selection**.

Birds' beaks are like tools – different ones are suited to different jobs. Beaks of different shapes are adapted to eating different kinds of food.

Try to match each beak with the food below.

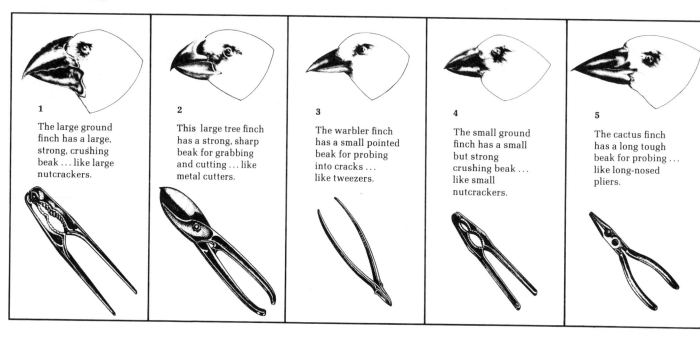

1
The large ground finch has a large, strong, crushing beak ... like large nutcrackers.

2
This large tree finch has a strong, sharp beak for grabbing and cutting ... like metal cutters.

3
The warbler finch has a small pointed beak for probing into cracks ... like tweezers.

4
The small ground finch has a small but strong crushing beak ... like small nutcrackers.

5
The cactus finch has a long tough beak for probing ... like long-nosed pliers.

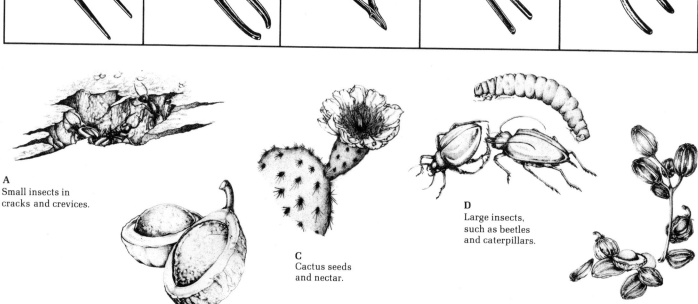

A
Small insects in cracks and crevices.

B
Large, hard seeds.

C
Cactus seeds and nectar.

D
Large insects, such as beetles and caterpillars.

E
Small, hard seeds.

How were the finch species formed?

The shapes of the beaks of the Galapagos finches can be explained by natural selection. But natural selection is only part of the story. The geography of the Galapagos Islands and the surrounding ocean must have played an important part in the formation of the many finch species that we see today.

The diagrams below show how the finch species may have evolved.

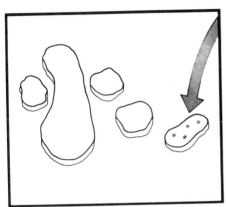

1 At first there were no finches on the Galapagos Islands. Then some finches from the mainland somehow managed to reach one of the islands.

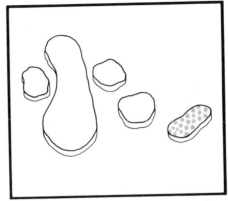

2 The finches increased in numbers and, under the influence of natural selection, gradually became adapted to the local environment.

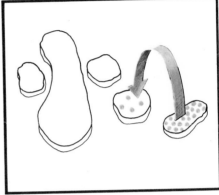

3 Some of the finches managed to fly to a second island, where the environment was different.

4 The finches on the second island gradually became adapted to their new environment.

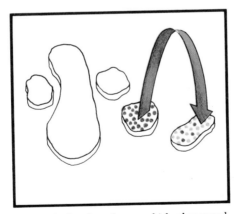

5 Some finches from the second island managed to fly back to the first island. But they had become so different from the finches already there that they could not interbreed with them. The two populations had become **two different species**.

6 This process was repeated over and over again as eventually the finches colonized the other islands. Now there are 13 different finch species on the Galapagos Islands.

Answers to p. 112
Beak 1 is adapted to eating food B
Beak 2 is adapted to eating food D
Beak 3 is adapted to eating food A
Beak 4 is adapted to eating food E
Beak 5 is adapted to eating food C

The Galapagos finches are generally regarded as one of the classic examples of evolution involving natural selection.

Are there any other examples?

The cichlid fishes of Lake Victoria

Natural selection can be used to explain the evolution of many groups of species in various parts of the world. One group of particular interest to scientists here at the Natural History Museum in London is the **cichlid fishes** that inhabit Lake Victoria in East Africa.

Lake Victoria contains more than 200 very similar species of cichlid fishes. The different species vary slightly in colour, size and shape, but the main differences between them are in their jaws and teeth – which are adapted to eating different kinds of food. Some eat other fishes, some eat insects, others feed on plants. Between them, the fishes eat almost every kind of food in the lake.

How were all the species formed?

A closer look at Lake Victoria

The history of Lake Victoria provides a number of clues to the origin of the different species . . .

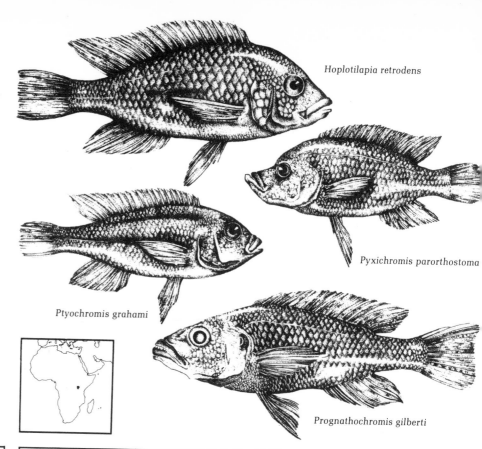

Hoplotilapia retrodens

Pyxichromis parorthostoma

Ptyochromis grahami

Prognathochromis gilberti

Lake Victoria probably began to form about a million years ago, when earth movements created a number of small lakes along the existing rivers.

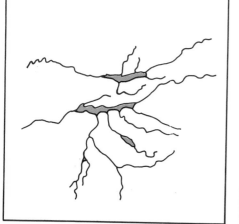

Gradually, the lakes grew larger, and some of them merged.

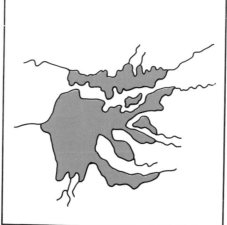

Further earth movements occurred. Droughts caused the lakes to shrink and split, and floods caused them to grow and merge again. Over the years, this process seems to have been repeated a number of times, until eventually the present-day lake was formed.

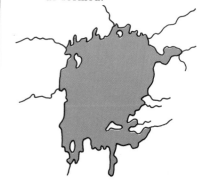

How were the fish species formed?

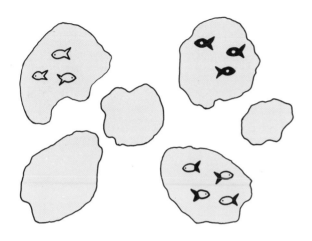

1 At first there were probably only one or two species of cichlid fishes living in the rivers and the newly-formed lakes.

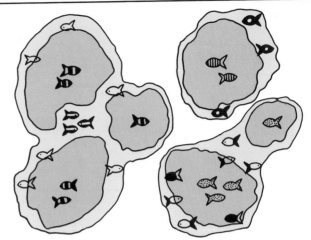

2 As the lakes grew larger, new habitats were formed – swamps in the small shallow lakes, for example, and deep water in the large lakes. Each new habitat provided new kinds of food. Over many generations, under the influence of natural selection, groups of fishes gradually became adapted to these new habitats and new kinds of food.

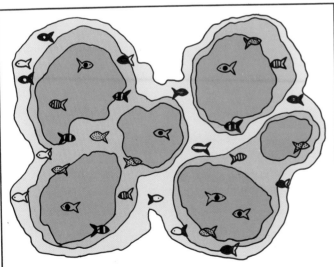

3 Later, when the lakes merged, the groups of fishes came together. But they had become so different that they could no longer interbreed – they were **different species**.

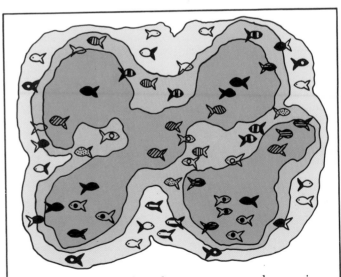

4 Since then, there have been many more changes in the lakes, and many more opportunities for new species to evolve. Now there are more than 200 species of cichlid fishes in Lake Victoria – each one slightly different from the others.

A last word

Darwin's theory of evolution by natural selection can be used to explain the origins of many living species. The Galapagos finches and the cichlid fishes of Lake Victoria fit the theory. But these are only a tiny fraction of the countless numbers of species alive today. What about the others?

In many cases, it is difficult to apply the theory because not enough is known about the organisms, their history or their environment. In other cases, the theory doesn't seem to explain the situation – at least not very well. Does that mean that the theory of evolution by natural selection is wrong?

Ever since Darwin published *On the Origin of Species,* his theory has been debated and discussed by scientists, theologians and philosophers – and the debate still goes on. Over the years a great deal has been written on the subject and, for those who would like to know more about the arguments, some references have been included in the booklist.

Natural selection

In this book we have shown that natural selection is a very simple theory. It follows logically from four basic ideas – each of which can be illustrated by examples from the natural world. Natural selection does, in fact, fit in with what we observe in nature, and can be seen in action in living populations.

But can natural selection explain the evolution of new species? Darwin thought that it could, and it certainly seems to offer a convincing explanation for the origin of species such as the Galapagos finches.

Other theories

Does natural selection offer the only explanation of evolutionary change?

Today it is generally accepted that other mechanisms, some not yet fully understood, may also have played their part in the evolution of new species. Since Darwin's time, evolutionary theory has been expanded and modified, with new evidence continually being added from molecular biology, population dynamics and, in particular, from genetics.

The theory of natural selection and the debate surrounding it have stimulated an enormous amount of research and raised a great many questions:

What factors influence the changing patterns of variation amongst living things?

What role can fossils play in helping us to interpret evolutionary change?

Does chance play an important part in evolution?

These are just a few of the questions still to be answered. But the theory of natural selection remains central to any study of evolution, and is one of the keys to our understanding of the diversity of life.

Further reading

... about Charles Darwin

The autobiography of Charles Darwin, 1809–1882, edited by Nora Barlow, Collins, 1969. A fascinating book, written for his own children.

Charles Darwin by John Chancellor, Weidenfeld & Nicolson, 1974. An illustrated biography.

Darwin and the Beagle by Alan Moorehead, Penguin Books, 1979. A well-illustrated account of the voyage of the *Beagle* and the impact it had on Charles Darwin and on the society in which he lived.

Down House, at Downe near Bromley in Kent, was Darwin's home for almost 50 years, and still contains many of his books, papers and personal belongings. It is in the care of the Royal College of Surgeons, and can be visited by members of the public (telephone Farnborough 59119 for details of opening times).

... about natural selection and evolution

On the Origin of Species . . . by Charles Darwin, 1859. Available today in paperback from Penguin Books. The original, well-reasoned argument for natural selection.

The Illustrated Origin of Species by Charles Darwin, abridged and introduced by Richard Leakey, Faber & Faber, 1979. A new, shorter and illustrated version of Darwin's classic book.

Evolution by Colin Patterson, British Museum (Natural History), 1978. An authoritative introduction for the general reader. It includes sections on genetics and some of the philosophical aspects of modern evolutionary theory, and has a useful reading list.

The Story of Evolution by Ron Taylor, Ward Lock, 1980. A colourful and easy-to-read book that covers a range of evolutionary topics from the origin of life to genetic engineering.

Life on Earth by David Attenborough, Collins/BBC, 1979. A vivid account of the history of life. A new and even more extensively illustrated edition is now available, published by Reader's Digest.

Evolution by T. Dobzhansky *et al.*, W. H. Freeman, 1977. A comprehensive account for the student and specialist.

Modes of Speciation by Michael J. D. White, W. H. Freeman, 1978. A thorough discussion of the different ways in which new species may be formed. Includes a chapter on polyploidy.

Darwin Retried by Norman Macbeth, Gambit paperback, 1979. A lawyer's view of the evidence for natural selection.

... to keep up to date

The best places to find out about the latest ideas on natural selection and evolution are the biological magazines and journals. *New Scientist* and *Scientific American* are two of the most readily available, and regularly contain articles written by leading evolutionary biologists.

Index

Acknowledgements

10: portrait of Charles Darwin by George Richmond, by courtesy of George Darwin Esq.
23: portrait of Alfred Rendle, from *Journal of Botany*, vol. 76, 1938.
66/67: Darwin manuscript, by courtesy of Down House and the Royal College of Surgeons.
81: infant with sickle-cell anaemia, by courtesy of Professor P. D. Marsden, University of Brazil.

Photographs

24/25: gannets, *A. Winspear Cundall/ Natural Science Photos.*
30: foxgloves in the wild, *G. Maclean/ Oxford Scientific Films*; king penguins in the wild, *Dr Giles Clarke.*
33: king crabs, *David Attenborough.*
45, 49: red deer stags fighting, *Hans Reinhard/Bruce Coleman Ltd.*
49: rabbit, *Jane Burton/Bruce Coleman Ltd.*
58: cuckoos, *Ian Wyllie.*
76: nuthatch eating peppered moth, *BBC Natural History Unit*; Manchester factory, *Northern Picture Library*; insecticide spraying, *B. Brooks/Bruce Coleman Ltd.*
77: 1940s crowd, *BBC Hulton Picture Library.*
79: Sierra Nevada, *M.P.L. Fogden/ Oxford Scientific Films.*
81: male and female golden pheasants, *Eric and David Hosking*; pheasants fighting, *Bruce Coleman/Bruce Coleman Ltd.*
97: sticky potentilla, *Jepson Herbarium, University of California.*
108/109: Galapagos Islands, *Professor Robert I. Bowman.*

For their cooperation and help in the production of the following photographs, we should like to thank:
8, 9: Darwin's study, Down House and the Royal College of Surgeons.
20: football crowd, Chelsea Football Club.
42: oak woodland, Selborne Society.